THE

'In this book, Paul ⬚⬚⬚⬚⬚ It has to be one of th⬚⬚⬚⬚⬚ ⬚- logy-focussed book⬚ ⬚⬚⬚ year. Brown has an excellent journalistic style, and pulls the reader on relentlessly through the tales of technical inspiration and human weakness that litter the history of the rocketbelt. The book overall is a delight to read, and the story is genuinely one where reality is stranger than fiction. Recommended.'
- Brian Clegg, *Popular Science .co.uk*

'This diverting story begins with the history of the rocketbelt, from Buck Rogers' fantasy to the real personal flying machine developed for the US Military. It reads like good movie material.'
- David Hambling, *BBC Focus*

'The Rocketbelt Caper begins with a brief history of rocketbelts that's probably the best ever in print.'
– Bill Higgins, *Beamjockey*

'This book is a must-have. Great research on the secret history of rocketbelts. Recommended.'
- Peter Gijsberts, *Airwalker Society of Rocketbelt Enthusiasts*

As seen in *New Scientist* and *Popular Science*.

THE ROCKETBELT CAPER

A True Tale of Invention, Obsession and Murder

PAUL BROWN

tonto press
www.tontopress.com

Published in the UK in 2007 by Tonto Press
Copyright © Paul Brown 2007
All rights reserved
The moral rights of the author have been asserted

The names of some of the characters
in this book have been changed

ISBN-13:
9780955218378

British Library Cataloguing in Publication Data:
A catalogue record for this book is available
from the British Library

First published in the US in 2005

Printed and bound at Athenaeum Press Ltd,
Gateshead, Tyne & Wear

Cover illustration by Garen Ewing

Tonto Press
Newcastle upon Tyne
United Kingdom

www.tontopress.com
www.rocketbeltcaper.co.uk

CONTENTS

PROLOGUE

It was just past midday on a January Sunday in 1995, and Brad Barker was getting nervous. It was warm and dry, but grey clouds were gathering overhead. A light breeze whipped at the long grass that ran along the edges of the runway. The water in the airport's sea-lane rippled and mirrored the dull sky. Barker strode out to the runway in white sneakers, jeans and a denim shirt, a mobile phone clipped to his belt, and a pair of unnecessary sunglasses perched on his nose. Around him, a handful of colleagues busied themselves with preparations. This was supposed to be the beginning of the dream, but just one mistake could turn it into a nightmare. And what the hell would he do then?

He'd tested it, of course, as much as you could without strapping a man into it and blasting him into the air. Now it was time to do just that. Barker looked at his pilot, bespectacled and slightly out of shape, wearing a white helmet and a jumpsuit. Barker was sure he had the right man for the job. Bill Suitor was nothing if not experienced. He'd flown these things hundreds of times before. Suitor wouldn't have agreed to do this unless he thought it would work. And there was no reason to think it wouldn't. That's what Barker told himself as he strapped Suitor into the Rocketbelt 2000.

It had taken five long years and several hundred thousand dollars to reach this point. Was it an obsession? You could say so, as the idea of building a rocketbelt had dominated Barker's thoughts ever since he first saw the amazing device, in a James Bond movie, as a nine-year-old boy. There had been sacrifices made, friendships lost, legal wrangles, and a bunch of other stuff he didn't want to think about right now. It had been quite an ordeal, but that was all in the past. Barker slapped Suitor on the back and retreated to a safe distance.

The pilot offered a thumbs-up, and there was a silent moment of anticipation. Barker studied his 'Pretty Bird' - its lovingly polished fuel tanks, curved exhaust pipes and control handlebars he'd worked so hard to put together. It was a unique device, but it looked oddly familiar, being as it was the realisation of a childhood fantasy. Then Suitor twisted open the throttle. The test flight began.

It was the noise that hit the small band of onlookers first. A high-pitched wail – an explosive scream of superheated steam, as loud as a jet engine. It was an assault on the eardrums, like having an aerosol fired into each ear. And this from a device not much bigger than a portable vacuum cleaner, strapped to the pilot's back like a hiking rucksack.

A white cloud of steam erupted from the exhausts, kicking up a swirl of dust around the pilot's feet. And then – in a defining moment for Barker – pilot and machine lifted into the air. There were no wings and no strings. This man was flying solely on the power of the rocketbelt.

Suitor hovered for a moment, a couple of feet from the runway, then manoeuvred out over the sealane, blasting a spray of water into the air. He banked into a graceful circular flight path, increasing his altitude to ten, fifteen, thirty feet. Barker watched the figure of the rocketbelt pilot silhouetted against the sky, his shadow becoming increasingly longer on the runway below. It worked. It really worked. And all Barker could think was, when do I get a turn?

Fuel was limited, and the flight needed to be short. So Suitor arced back towards his starting point, and began his descent. The landing was feather-light and perfect. The pilot bounced on tiptoes, steadied his feet, then offered Barker a salute.

'Yeah!' whooped Barker. He rushed over to congratulate the grinning Suitor, and said, 'Let's fuel it up and go again.'

So they did, and the second test flight was just as successful as the first.

Then, on the third flight, things began to go wrong. In fact, if the truth were known, things had been going wrong for quite some time. They would get a lot worse before this whole sorry affair was over. And it was all because of one of the most sought after and revered machines in the history of flight. It was all because of the amazing rocketbelt.

Blame Buck Rogers, and James Bond, and Commando Cody, and the Jetsons. It was their fault that everybody wanted a rocketbelt. The iconic device was a ubiquitous feature in twentieth century science fiction. It became pivotal to the accepted vision of the future. Eminent science writers like Isaac Asimov confidently predicted that, by the year 2000, the citizens of Earth would be zipping about through the air using rocketbelts. That never quite came to pass. Brilliant inventors did build a small number of working rocketbelts, and a handful of pioneering pilots got to fly them through the sky. But, for most of those who coveted rocketbelts, the device remained tantalisingly out of reach, a technological holy grail.

Then Brad Barker and two of his friends got sick of waiting for science fiction to become reality. They formed a fractious partnership and set out on an ambitious quest to build their very own rocketbelt. Perhaps the least incredible thing about their story is that they actually succeeded in building and flying a working rocketbelt – in fact the very best rocketbelt that had ever been built. Then the incredible dream turned into an almost unbelievable nightmare.

For now, Barker was watching his rocketbelt fly through the air, grey skies closing in, wind brushing the back of his neck. Suitor circled over the sea-lane, and swung back around to the runway, but he came in too fast. Much too fast. His feet clipped the tarmac and flipped him around, hard onto the ground. The rocketbelt hit the runway at pace, too, with a sound like a car hitting a lamppost. The pilot cut the throttle and lay still on the tarmac. Miraculously, Suitor was unhurt, if a little shaken. Barker rushed onto the runway to check out the damage.

'The thing damn near killed me,' said Suitor, crawling to his knees and unstrapping himself from the device.

'You damn near killed my rocketbelt,' said Barker. He ran his hands along the dented stainless steel he'd spent so long polishing. Goddamn it, this would take months to repair. And it was not as if Barker didn't already have enough problems to deal with.

A few spots of rain began to land on the tarmac. Barker hoisted up the rocketbelt and loaded it into his trailer. He would have to put his plans for a public demonstration flight on hold for a while. Then, once the belt was repaired and public flights were arranged, the money would begin to roll in. Finally, he would be able to forget about his troubles. That's what Brad Barker reckoned. In fact, Brad Barker's troubles had barely even started.

ONE
FROM SCI-FI TO REALITY

In 1928, Buck Rogers fell asleep and woke up in the 25th century. Anthony 'Buck' Rogers was a former US Air Force pilot who served over France in the First World War. Returning to the US, Buck began to work as a surveyor in Pennsylvania. Then, on that fateful day in 1928, he was caught in a cave-in while inspecting a mine. Trapped underground, Buck was overcome by a strange radioactive gas and passed out. When he awoke, it was the year 2419.

The story of Buck Rogers first appeared in the August 1928 edition of pulp science fiction magazine *Amazing Stories*, which published a novella by Philip Francis Nowlan entitled *Armageddon 2419 AD*. In the novella, Buck wakes to find the Earth under siege by the deadly Mongols. Buck is made captain of the planet's defensive forces – the Rocket Rangers. He lives in a city made of 'metalloglass', eats synthetic food, uses a ray gun to fight the Mongols, and – most excitingly – travels through the air using a flying backpack, originally called the 'jumping belt'.

The success of the *Armageddon 2419 AD* novella led to author Phil Nowlan being commissioned by the National Newspaper Syndication Service to write the world's first science fiction comic strip. *Buck Rogers in the 25th Century* debuted in 1929 and was subsequently published in over four hundred newspapers and translated into eighteen languages.

But the definitive Buck Rogers realisation was the 1939 movie serial starring Larry 'Buster' Crabbe, and it was this cliffhanger-filled adventure that really introduced the public to the concept of a flying backpack. Among the art deco sets, sparking ray guns and fantastical space ships, the embryonic rocketbelt – named in the movie serial as the 'de-gravity belt' – very much stood out. The device officially became known as a 'rocketbelt' after appearing in a spin-off newspaper adventure strip about Buck's buddy, Buddy.

After the success of the Buck Rogers filmstrip, the rocketbelt returned to cinema screens in 1949 in the twelve-episode Republic Pictures serial *King of the Rocketmen*. The serial starred Tristram Coffin as Jeffrey King, a rocket propulsion expert who dons an atomic rocket pack and bullet-shaped helmet to fight the evil Doctor Vulcan as 'the Rocketman'. *King of the Rocketmen*'s thrilling flying scenes were filmed using an oversized dummy that was slid along wires strung out across the Hollywood hills. The flying footage was so effective and popular that it was recycled and used again in the 1950s in a set of Republic serials featuring a new Rocketman – Commando Cody.

The first Cody serial, 1952's *Radar Men from the Moon*, saw actor George Wallace don the rocket pack and helmet to battle a mysterious race of 'Moon Men'. In the serial, Cody uses his amazing flying backpack to avoid ray guns, survive bomb blasts, and escape from volcanic eruptions, before eventually defeating the lunar bad guys. A further Commando Cody serial, *Zombies of the Stratosphere,* was released later in 1952, and then a TV show, *Commando Cody: Sky Marshall of the Universe*, was screened in 1955. Oddly, despite the TV show's title, the character of Cody didn't feature in the series at all. The new hero, played by Judd Holdren, was called Larry Martin, and he wore a Zorro-style mask rather than Cody's bullet-shaped helmet.

The rocketbelt would go on to feature in dozens of other movies and TV shows, most notably *Thunderball*, but also *Lost In Space, Ark II, The A-Team,*

Minority Report, and the 1991 Disney movie *The Rocketeer*, which revisited many of the themes of the original Rocketmen serials. Set in 1930s Hollywood, *The Rocketeer* stars Bill Campbell as Cliff Secord, a test pilot who discovers a top-secret rocketbelt, designed by eccentric billionaire aviator Howard Hughes, and uses it to battle Nazi spies and save love interest Jennifer Connelly. Superb retro design and a memorable rocketbelt-powered fight scene on top of an airship helped make *The Rocketeer* a cult favourite. But the rocketbelt remained indelibly linked to Buck Rogers, having fuelled a generation of boyhood dreams. It would be Rogers that real-life rocketbelt pilots would later point to as their inspiration.

Science fiction had introduced the public to the possibilities of a portable device that would allow the wearer to fly through the air. The rocketbelt became synonymous with a golden vision of the future, and visualisations of the world of tomorrow invariably featured silver-suited citizens zooming about a *Jetsons*-style city wearing flying backpacks. Futurists began to predict real developments in rocketbelt technology.

In 1965, respected sci-fi writer Isaac Asimov wrote in *Science Digest* of a rocketbelt that could 'lift a man clear off the ground.' He predicted that, by the year 1990, rocketbelts would be practical and cheap enough to use for short trips, and imagined office workers soaring above traffic jams on their way to work. 'This will be a thrill in the first place – who has not dreamt of flying?' Asimov wrote. 'Watch out for engine failure, however.'

Even before Asimov made that bold prediction, thanks in no small part to Buck Rogers and his sci-fi contemporaries, inventors around the world had become intrigued by the concept of a flying back-pack. The public – and the Army – saw potential in a portable rocketbelt for everyday and military use. A working rocketbelt could revolutionise travel and warfare. In truth, it could revolutionise modern life. Whoever could invent such a device would surely become world famous and seriously rich. Now all that remained was for someone to turn the science fiction into science fact.

But had a serious effort to make a real rocket-powered personal flying device already been made? Although the details remain shrouded in contro-versy, an attempt was allegedly made by the Nazis towards the end of the Second World War. Nazi scientists, boasting the expertise of rocket expert Wernher von Braun, had abandoned dreams of space travel, and instead attempted to tackle the problem of moving infantry across battlefields quickly, avoiding obstacles such as minefields and barbed wire. It has been suggested that Von Braun, an honorary Major in the SS, may have personally overseen the project at the Peenemünde rocket research centre, where he had developed the V-2 liquid fuel rocket and other devastating weapons.

The proposed rocket pack – the Himmelstürmer, or Skystormer – used two low-powered rockets, one strapped to the pilot's chest, and the other to his back. The second rocket ran at a slightly higher velocity, producing forward motion. The packs were said to enable leaps of up to 180 feet, and were apparently cheap to run. But, if they did exist, they were never used in military action.

The Peenemünde plant was destroyed towards the end of the war, before the advancing Russian army could reach it, and no trace of the project was ever officially found. Although a lack of evidence suggests the Nazi rocket packs may have been pure propaganda, some enthusiasts claim the packs were successfully built, and tested by a secret Nazi storm-trooper unit. Nazi scientists were undoubtedly responsible for a large number of technological marvels during the war, including military jets and rudimentary cruise missiles. Many of their projects were deliberately obliterated before the Allies could get their hands on them. But could the Skystormer, or its plans, have been appropriated by the US Army?

At the end of the war, Wernher von Braun and approximately one hundred other Peenemünde scientists were brought to the US as part of Operation Paperclip. These men played a significant role in the early development of the US space program. Von Braun became a US citizen in 1955, despite his association with the Nazi rocket development program, during which over twenty thousand concentration camp inmates and Allied prisoners of war were said to have died of starvation and mistreatment. Von Braun later worked for the Walt Disney Corporation, designing theme park rides, working with feature film animators, and hosting television specials, including *Man and the Moon* – a 1955 dramatisation of a moon landing bearing remarkable similarities to the real 1969 landing. Prior to that, von Braun and his colleagues were located at Redstone Arsenal military base in Alabama during the development of the first properly-documented flying rocket pack.

It was 1952, and the man behind the rocket pack project was civilian radar technician Thomas Moore. He built the pack at the Redstone base using a US Army grant of $25,000. Moore's rocket pack was called the Jet Vest, and the inventor tested it himself, flying it while tethered into a metal frame. But the Army funding ran out before the project could be completed. Moore's almost-realised dream collapsed, although he continued to develop his idea on paper, and filed further patent designs up until 1966.

In 1958, Harry Burdett Jnr and Alex Bohr, engineers from the Thiokol Corporation's Reaction Motors Division plant at Bingham City, Utah, embarked upon Project Grasshopper for the US Army. Their goal was to develop a rocket pack that would allow infantrymen to leap great distances and run at high speeds. Versions of the Thiokol Jump Belt used bottled nitrogen, and then rockets, for thrust. They apparently allowed the user to leap up to 25 feet into the air and sprint at up to 30 miles per hour. The Jump Belt was demonstrated in front of military top brass at Fort Bragg, North Carolina but, again, funding dried up and the project was shelved.

But, within a few years, the US Army would have in their possession a working rocketbelt. Science fiction was becoming reality, and man's dream of a personal flying device was about to come true.

TWO
THE BELL ROCKETBELT

The Edwards Air Force Base is located in the middle of the Mojave desert in Southern California, hundreds of miles from anywhere else, and surrounded by nothing but sand and dry lake beds. In 1947, then called the Muroc Base, it was the perfect place for the Bell Aircraft Company to test their radical new supersonic jet – the X-1. The bullet-shaped, bright orange X-1 was powered by four rocket engines, and built to withstand 18-times the force of gravity.

The desert test flights were designed to push the jet up to and past Mach 1, a velocity of approximately 760 miles per hour – the speed of sound. Over several months from July 1947, Captain Chuck

Yeager pushed the X-1 nearer and nearer to the target velocity. The jet was launched from the bomb bay of a B-29 Superfortress and glided free for a few moments before its rocket engines were fired. Its heavy fuel consumption allowed the rockets to run for just a few minutes, after which the X-1 glided to a landing on a nearby dry lake. Each flight saw the speed of the jet increased by only a tiny increment in order to maximise safety.

By the time of the ninth flight, on the morning of 14 October 1947, the X-1 team was ready to push for Mach .98. The jet was launched as normal, and Yeager pushed toward his target speed. But, at Mach .965, the metre oscillated wildly and shot off the scale. At exactly 10.18am, the plane disappeared from view. Ground staff heard a huge noise like a thunderclap. At first, they feared Yeager and the X-1 had been lost. In fact, what they had heard was the world's first sonic boom. Yeager had broken the sound barrier. For 20 seconds the X-1 travelled faster than the speed of sound – it was the world's first supersonic jet, and Yeager was the fastest man in the world.

Key components of the X-1 were reaction control thrusters. These miniature rocket engines, mounted on the tail and wingtips, helped the pilot control the jet at high altitudes, where the thin air hampered aerodynamic flight characteristics. The thrusters have since been used to control manoeuvrability during space exploration. They were invented by Bell propulsion engineer Wendell Moore, from Canton, Ohio. Moore joined Bell in 1945 and later became the company's assistant chief engineer. (The Bell Aircraft Company, founded in 1930 by Lawrence Bell, later became Bell Aerosystems, and then Bell Aerospace.)

In 1953, while working on the X-plane project, Moore considered strapping his miniature rockets to a human being. 'I could stick two of those on a man's back and make him fly like Buck Rogers,' he told a colleague. Moore explained his idea by drawing a plan in the desert sand with a pointed stick, and then sketched a design on a pad showing valves and gauges and high-pressure lines, and a man with a rocket pack attached to his back.

Moore called his invention the Small Rocket Lift Device (SRLD). The basic principle of the design involved gas being forced at high pressure through downward-facing exhaust tubes to generate lift. In order to test his idea, he built a nitrogen-powered rig manufactured from steel tubing, with two exhaust tubes fitted with rocket nozzles. The device was tethered to safety cables, and fed by a high-pressure hose, with gas flow controlled from the ground by a technician.

Moore tested the nitrogen rig himself, inside a large Bell hangar. At the first test flight on 17 December 1957, the 54th anniversary of the Wright brother's first flight, Moore, with a helmet pushed over his typical 1950s flattop haircut, offered a thumbs-up signal to a fellow engineer, who opened up the thrust control.

With his Bell colleagues holding onto the safety tethers, Moore lifted into the air and hovered four inches from the ground, wobbling around in a small circle. Further test flights pushed the ability of the device. Refinements were made with the help of Moore's colleague Jim Powell and Bell cameraman Tom Lennon, both of whom tried out the rig to help iron out stability problems. It was Lennon who spotted that, in order to keep the device under control, the pilot should keep his feet together. By 1959, Moore had achieved a stable flight 15 feet above the ground.

Meanwhile, the US Army's Transport Research and Engineering Command (TRECOM) was becoming very keen on developing a working rocket pack. Having been closely involved with Thiokol's aborted Project Grasshopper, TRECOM turned to the Aerojet General Corporation, another company with experience in rocket propulsion. In 1960, Aerojet engineer Richard Peoples made a successful tethered test flight of their device – the Aeropak. But the Aeropak project stalled, and TRECOM's attentions shifted to Bell and Wendell Moore.

TRECOM contracted Bell to 'build a Small Rocket Lift Device and demonstrate its feasibility in manned free flight.' Wendell Moore was to be Bell's technical director for the project. The budget was set at just $150,000, so Moore recycled parts from previous aircraft and spacecraft, like the Mercury space capsules. Working at the Bell Aerosystems plant in Buffalo, New York, Moore radically improved his design. He also renamed the device. Now the SRLD became a 'Rocketbelt'.

The Bell Rocketbelt consists of three tanks fitted to a form-fitting fibreglass corset. At the sides, two downward-facing pipes are connected to rocket nozzles. At the front, handle grip controls for throttle and steering are fitted to handlebars that run under the operator's arms.

The rocketbelt is powered by a throttleable rocket motor that runs on hydrogen peroxide – an alkaline chemical commonly used in bleaches, disinfectants, dyes and antiseptics. Hydrogen peroxide (H_2O_2) is a clear, dense liquid – basically water with an extra molecule of oxygen. Household bleaches and antiseptics contain around a three percent concentration of H_2O_2. Rocketbelt fuel contains around a 90 percent concentration. At such high concentration, and high volatility, H_2O_2 makes for a very effective – if very dangerous – rocket fuel.

On the rocketbelt, the middle tank contains nitrogen, and the outer tanks contain H_2O_2. The nitrogen acts as a pressurising gas. A valve is opened, and the nitrogen is released into the fuel tanks, forcing the hydrogen peroxide at high pressure over a catalyst bed of fine-mesh silver screens. A reaction takes place and steam is produced. The steam is then fired out of the rocket nozzles at high velocity, creating thrust, and allowing the rocketbelt to fly.

The first rocketbelt test flight took place on 29 December 1960. Tethered into a safety harness, Moore piloted the rocketbelt into the air for several erratic seconds and made a series of rudimentary manoeuvres using the new control system. Over a succession of further tethered tests, the Bell team troubleshot and gradually improved the rocketbelt. Then, on the 20th test flight, something went wrong.

During a flight in February 1961, the rocketbelt snagged on a safety tether. Moore was unaware of the mishap until the tether snapped, and both pilot and belt plunged eight feet to the ground. Moore badly fractured his knee and would never fly his rocketbelt again. It would be another Bell employee who would attempt the first free flight.

Harold 'Hal' Graham was a 27-year-old science graduate from Buffalo. He had been working for Bell as a test engineer for just over a year when he was selected to pilot the rocketbelt. It would be Graham's first taste of flying. He was not a registered pilot, and the only machine he had previous experience of driving was a car. He was, however, a rocketbelt fan, having grown up with Buck Rogers comics and Commando Cody serials. When Bell began to ask around for a volunteer to fly the rocketbelt he had no hesitation in applying for the job.

Graham's first tethered flight took place two weeks after Moore's accident, in March 1961. These flights took place in a large aircraft hangar. The rocketbelt was suspended from the ceiling, and small amounts of thrust were used to generate moderate lift. 36 tethered flights later, it was time for the safety ropes to come off.

The very first untethered rocketbelt flight took place at seven in the morning on 20 April 1961. A 20-man Bell crew gathered at an empty clearing near the Bell plant on Buffalo's Niagara Falls Boulevard and opposite the Niagara Falls Municipal Airport, which had been specially closed for 30 minutes. The crew ran through a detailed checklist in preparation for the flight. Then Graham, wearing a black rubber suit, white helmet, work boots, and goggles, released the throttle in a short burst to check the propulsion. All seemed fine. Again he released the throttle, this time successfully lifting the belt around 18 inches from the ground in a thick cloud of steam, and piloted it in a straight line at a speed of around ten miles per hour. The noise was incredible – an explosive roar of gas as loud as a pneumatic drill. And visibility was poor – almost zero according to Graham – due to condensation created by the rocket exhaust. On the first free rocketbelt flight Hal Graham flew for 13 seconds and covered a distance of 112 feet – eight feet less than the Wright Brothers had covered in their inaugural flight. It was nevertheless a thoroughly triumphant debut.

Following the success of the test flight, Bell executives were keen to unveil the remarkable device to the public. But Wendell Moore was insistent that there should be many more tests before that happened. He wanted to iron out safety and reliability problems before any public demonstration. So a succession of secret test flights were performed at the Niagara Falls Airport, and on a golf course at the Youngstown Country Club in New York. The tests saw Hal Graham perform turns, negotiate obstacles and pilot the belt over hills and streams. After 28 free flights, Moore was satisfied enough to agree to a public demonstration.

The first public rocketbelt flight took place at Fort Eustis, Virginia, on 8 June 1961 at a TRECOM demonstration of new technologies. TRECOM officials and scores of press reporters gathered on the Fort Eustis runway as the Bell crew helped Graham get kitted out in his protective suit and fitted into the belt. Because the rocketbelt corset was rigid and form-fitting it required something of a squeeze to get Graham into it. Safety checks were performed, and the crowd stood back. Then Graham took off with a huge roar. Light bulbs flashed and film reels rolled as Graham piloted the rocketbelt into the air, legs swinging below him. Conditions were good, and the steam from the exhaust stirred up only a light cloud of dust from the runway. Against a backdrop of Air Force planes, Graham manoeuvred the rocketbelt over a truck, and higher into the sky. He flew to around 15 feet, and then descended, bouncing slightly as he landed on his feet. Graham then offered a salute.

After removing his fire suit, Graham was mobbed by the press. Microphones were thrust into his face, and pencils jotted down every word he said. Bell officials handed out press releases which began, 'Harold M Graham is believed to be the first man to fly with back-carried rocket equipment.'

The story made the front pages across the US. The *New York Times* headline read, 'Portable army rocket propels man 150 feet in 11-second test flight.' *Life* magazine said, 'Graham was strapped to a hydrogen peroxide-fuelled rocket. The Army hopes it will someday make all foot soldiers look like Buck Rogers.'

One week later, Graham demonstrated the rocketbelt on the front lawn of the Pentagon in Washington DC in front of a huge crowd of military personnel. An estimated 3,000 Pentagon staff left their desks to view the demonstration. In the first of two flights, Graham took off in a swirl of blades of grass, negotiated a parked army sedan, and flew for almost

150 feet before touching down safely.

Public reaction to the rocketbelt was stunning. It seemed everyone wanted to see the flying rocketman – a real life sci-fi hero with his amazing rocketbelt. In October 1961, Graham, Moore and the Bell crew travelled to Fort Bragg in North Carolina to participate in another military demonstration, this time as part of a display of combat readiness. The demonstration was performed in front of a notable guest of honour – President John F Kennedy.

Graham, wearing a US Army uniform, took off from an amphibious landing vehicle, flew across a pond in a spray of water, and landed 14 seconds later on a sandy embankment in front of JFK. Graham remembered to salute but forgot to depressurise the belt in the excitement of the moment, although he managed to remain on the ground. 'Mr Kennedy was described by an Army Officer sitting near him as "wide eyed and open mouthed, just like a kid",' reported the *Buffalo Evening News*.

The public interest and publicity surrounding Graham and the rocketbelt generated much correspondence. Letters requesting public appearances began to flood the Bell offices. One man wrote to Bell requesting the use of the rocketbelt in order to claim a $1 million treasure trove that, he claimed, he could only reach with the use of the belt. Suspicious Bell executives turned the request down.

Although Hal Graham could now proficiently fly

the rocketbelt, he was still not a registered pilot. In November 1961 he decided to do something about that. He began to take flying lessons, and qualified for his pilot's license in July 1962. That year also saw the debut of the B-Series rocketbelt. The new belt was engineered to reduce weight, and, as rocketbelt pilot, Graham was kitted out in a brand new bright yellow flight suit.

But Hal Graham's short career as a rocketbelt pilot was coming to an end. During an ill-fated demonstration at Cape Canaveral, Graham fell 22 feet, landed on his head, and was knocked unconscious. He survived the crash, but decided to get out of the rocketbelt business. Graham made 83 untethered rocketbelt flights during his time at Bell, but he left the company in 1962 to pursue his new love of flying traditional aircraft. He set up his own one-man, one-plane charter flight company in Crossville, Tennessee.

After Graham departed, a new team of pilots comprising Robert Courter Jr, Peter Kedzierski, John Spencer and Gordon Yaeger flew the Bell Rocketbelt. Courter was a former Air Force fighter pilot and a veteran of the Korean War, and Kedzierski was a 17-year-old high school graduate in technical engineering and a keen glider pilot. John Spencer was Bell's chief test pilot and had trained with Chuck Yeager, while Gordon Yaeger (not to be confused with Chuck Yeager) was a Bell technician and a father of six. The Bell team took the belt all over the US, to South America, and to Europe, and made demonstrations in front of hundreds of thousands of spectators.

Bell had proved the principle of a rocketbelt could work, but so far all of the rocketbelt pilots had been technicians or pilots. It needed to prove to the US Army that the belt could be piloted by non-technical soldiers with minimal training. Bell needed to find an average kid with no flight experience and teach him to fly the rocketbelt. That average kid was 19-year-old William P Suitor, and he was about to become the most famous rocketbelt pilot in the world – a real-life rocketman.

THREE
THE ROCKETMAN

His name is Bond, James Bond, and he is British Secret Service Agent 007, licensed to kill. He is in Paris, on her Majesty's secret service, to investigate the mysterious death of evil SPECTRE agent Jacques Boitier.

So begins *Thunderball*, the fourth James Bond movie, starring Sean Connery as the suave spy. In the movie, Bond attends Boitier's funeral and becomes suspicious of the dead man's weeping widow. He jumps into his Aston Martin DB5 and follows her to a French chateau. Gaining entry, he surprises the widow – and surprises the audience – by punching her in the face, and knocking off her wig. For the 'widow' is actually Monsieur Boitier in disguise. The dastardly bad guy has faked his own death. Bond and Boitier fight, and a gang of SPECTRE guards burst in. They chase Bond up to the roof of the chateau, but the super spy has a surprise for them.

Displaying incredible foresight, Bond has stashed a rocketbelt on the roof. 'No well-dressed man should be without one,' he quips. He straps on the belt, pulls on a helmet, and releases the throttle. Bond shoots into the air, dodging gunshots from his pursuers, flies over the chateau, and glides down to the ground. Then, with help from a handy Bond girl, he stores the belt in the trunk of the Aston Martin, jumps into the car, and speeds away to safety. An impeccable escape.

Although audiences were amazed by this stunning pre-title sequence, few imagined that the rocketbelt was anything more than a movie special effect. Close-ups showing Connery flying the device were mocked up using a blue screen and a projected background, but the remaining shots of the rocketbelt flying from the roof of the chateau to the ground were real, and were filmed using the Bell Rocketbelt and a pilot named Bill Suitor.

Rocketbelt inventor Wendell Moore lived in suburban Youngstown, New York, and it wasn't unusual for him to take the Bell belt home at weekends. There it caught the attention of neighbourhood teenager and lawn jockey Bill Suitor. The 19-

year-old was on the verge of dropping out of college to join the Army. But Wendell Moore had other ideas. Moore offered Suitor a chance to fly the rocketbelt with the words, 'Hey kid, do you want a job?'

'It helped that he was a good family friend, and he and my father were against me joining the Army just as Vietnam was beginning to heat up,' said Suitor. 'The contract between Bell and the US Army stated that Bell must take a young man of average draft age and prove that he could be trained to pilot the rocketbelt. I used to cut Wendell's lawn and clean his windows. Talk about being a lucky bastard in the right place at the right time. I was very physically fit, able to water-ski barefoot and do all kinds of tricks on skis, and he thought that with such good balance I was a natural. It seemed he was correct.'

Suitor was officially hired by Moore on 26 March 1964. He worked as a gopher for the Bell team, and also began to train alongside the other pilots.

'I was trained on a tether that prevented me from crashing, using a cable and pulley system,' he said. 'My first tethered flight was with low fuel pressure so as to not allow me to leave the ground, just push me about the hangar under the tether, but allow me to feel the control inputs and what they did.'

Suitor underwent four months of tethered training. He experienced one accident, but managed to avoid injury. 'A weld on the throttle handle broke and I had no way of controlling the rocket,' he explained. 'I was whizzed about the hangar on the safety cable like a balloon that you inflate and just let go of.'

Despite that mishap, Suitor managed to master the belt without too much difficulty. The rocketbelt is controlled by two grip handles fitted to the handlebars. The right grip controls the throttle, and is twisted open to increase power. The left grip controls the rudder, so is manoeuvred left and right to control yaw. Pressing down or pulling up on both grips changes the direction of the thrust tubes, and causes forward or backward movement. Tilting the shoulders manoeuvres the belt left and right.

'Once I was able to translate the length of the hanger, turn a one-eighty, and return and land I was then sent outdoors, with no tether, for free flight,' said Suitor. 'I took to it like a duck to water. My first free flight was pretty uneventful. Your mind is so loaded with stuff you don't really think about what is going on. You just react.'

Suitor began to travel with the Bell crew to perform demonstrations, but the future of the rocketbelt was in doubt. It was expensive to run, and heavy fuel consumption meant flight times and distances were extremely short. TRECOM remained unconvinced about the feasibility of the belt, and it soon became clear that it was not prepared to send any more funding Bell's way.

With the Army's interest dwindling, Bell was forced to rethink its development strategy, teaming up with promoter Clyde Baldschun with the hope of generating money from entertainment bookings. Baldschun contacted Bell after his teenage son spotted the rocketbelt in a magazine. He was in the business of booking entertainment for state fairs and other outdoor events, and it was obvious the rocketbelt would be a real showstopper.

The first commercial rocketbelt flight took place at the Calgary Stampede, a huge annual Canadian event, in 1962. The theme of the event was transportation, so the rocketbelt was the perfect main attraction. The Bell crew flew the belt every night for a week and were paid $25,000, from which Baldschun took a 20 percent commission. Clearly the rocketbelt was going to be a huge money-spinner. Baldschun continued to organise regular bookings, and the rocketbelt's stock grew in the eyes of promoters and the public.

In 1964, the Bell crew travelled to the New York World's Fair, held at Flushing Meadow, Long Island, behind gates bearing the message: 'Welcome To The World Of Tomorrow!' Attractions included the world's biggest Ferris wheel – shaped like a US Rubber car tire, a gigantic Chrysler combustion engine, and Disney's *Small World* ride, where hundreds of animatronic children sang, '*It's A Small*

World After All! And then there was the rocketbelt. The US was undergoing a technological boom, and was just a few years short of putting a man on the moon. A lot of what had previously been thought of as science fiction was now becoming science fact. As the rocketbelt flew through the air over Long Island it became synonymous with technological innovation. For thousands of World's Fair visitors, the dream of a personal flying belt was now very much a reality.

Now gearing up for regular flights, the Bell team implemented a new safety feature to protect its pilots. The team had calculated the rocketbelt could only run for 21 seconds because of fuel limitations. This meant it was imperative the pilot had his feet back on the ground before those 21 seconds elapsed, lest he fall from the sky. The original safety alarm comprised of a buzzer and flashing red light inside the pilot's helmet that advised the pilot when only half of the fuel remained and landing was required. However, because of the noise and dust clouds created by the rocketbelt, it was difficult for the pilots to hear the buzzer or see the light. Instead, Bell fitted a vibrating box to the back of the pilot's helmet. The box vibrated intermittently after ten seconds, then continuously after 15 seconds, providing an effective warning to the rocketbelt pilots that they should land before they fell.

Bill Suitor's first public flight was at the 1964

California State Fair in Sacramento stadium. Suitor was supposed to take off from the stadium's racetrack, buzz the grandstand, and land on a stage. It seemed simple enough, until they turned off the stadium lights. With a spotlight shining in his eyes, Suitor took off and roared towards the stage, using only the audience's camera flashes to guide him. But he approached the stage at a very low altitude, and accidentally buzzed the orchestra pit. The blast from the rocketbelt sent music stands and sheet music flying, and musicians running for cover. Suitor managed to land safely, albeit among a mess of papers, wires and microphone stands.

Another key demonstration took place at Disneyland, California, in 1965. Suitor, in his white overalls and helmet, took off on a roaring cloud of steam, and adults and children in Mickey Mouse hats covered their ears. The rocketbelt perfectly complemented Disney's *Tomorrowland* futuristic attraction. It was a piece of the future realised in the present.

'Flying the rocketbelt is unlike anything else,' said Suitor. 'Once you get to a point where it is as natural as riding a bicycle - *that* is when you really enjoy it. You are never free of anxiousness. If you are there is something wrong with you, or you are a liar! Once you open that throttle, the noise, the feeling of flight, pure flight, with no wings, no strings... You don't even see the machine, just the two handles, and, like a bike, you just do it – fly. You don't think about what you are doing. Or at least I didn't. The most exciting thing is seeing your shadow race beneath you over rooftops or on the ground. That's when you realise what you are really doing, when you are zooming along and buzz something at high speed. I love it.'

In March 1965, Suitor and Gordon Yaeger travelled to Paris to film the opening sequence of *Thunderball*. The film crew had already shot close-up footage of Sean Connery wearing a dummy belt in the previous month, and now needed to film the real rocketbelt flying from the chateau to the ground.

'I was a kid who had just turned 20,' said Suitor. 'I was the backup pilot to Gordon Yaeger, who in reality was the primary pilot to do all the flying. As the day of the filming arrived and we were ready to go, he took from his pocket a five Franc coin and said, "Call it kid," as he flipped it in the air. He won, but he said, "Okay, I'll go first, then you, and we'll alternate flights till this is done." Gordon was 41, old enough and wise enough to realise James Bond was gonna be big, and he was willing to share it with me. We both made three flights, and footage of both of us appears in the movie, yet nobody realises anyone other than me flew in the movie.'

Footage from the six flights was spliced together to create one scene. The loud whining sound of the rocketbelt was overdubbed with the much quieter sound of a CO_2 fire extinguisher. And the Sean Connery footage was reshot with the actor wearing a helmet, after Suitor and Yaeger refused to perform the stunt without head protection. To his regret, Suitor never got to meet Connery during his time at the set. The rocketbelt went on to make another appearance in the James Bond series. In *Die Another Day*, Pierce Brosnan's Bond finds the dummy rocketbelt in the workshop of John Cleese's Q and asks if it still works. The answer? Yes, of course it does.

As Bell increased the frequency of their demonstration flights, they began to hire and train new pilots to help out. None of the new guys quite made the grade. One, Donald Hettrick, made several public flights before a couple of unfortunate crash landings caused his contract to be terminated. Dennis Comerford, a former Army flight instructor and engineer, also lost his job after making a bad landing during a demonstration flight. Another new pilot, Adam 'Doug' Miklejohn, made only one public flight, in Barstow, Florida, in 1965. Bill Suitor was on hand to capture Miklejohn's only moment of glory.

'I filmed it with my 8mm movie camera,' said Suitor. 'Poor Doug. His one moment of fame, and my camera had a speck of dust in the shutter, so none of his flight was recorded!'

Suitor continued to fly the rocketbelt around the

world. He visited 42 US states, plus Australia, Austria, Canada, France, Germany, Italy, Mexico, New Zealand, South Africa, Venezuela, and more. In July 1966, Suitor and Gordon Yaeger flew the rocketbelt in the UK at the British Grand Prix at Brands Hatch. Jack Brabham won the Grand Prix and the drivers' championship that day, and spectators were also treated to a promotional drag race between the rocketbelt and a Grand Prix car. (The car won – just.) Suitor also flew the rocketbelt at the first American Football Superbowl in 1967, in which the Green Bay Packers beat the Kansas City Chiefs.

But one particularly memorable set of rocketbelt flights stood out for Suitor. 'While making a tactical applications film for the US Army I got to fly through trees and under bridges,' he said. 'I really enjoyed those days, with no restrictions. I crashed into the water under a bridge, went under, kept going, righted myself, came up, and continued to fly like a big-ass Polaris missile out of a submarine. All by the grace of God more than skill!'

Wendell Moore continued to further develop the rocketbelt, producing designs for several variations on the original. Most of the new designs never got further than the drawing board, but the Bell Flying Chair was built and tested. It was essentially a rocketbelt fitted to an Eames fibreglass office chair mounted on castors. Another more successful development was the Pogo – this time essentially a rocketbelt fitted to the centre of a platform on which the pilot (or pilots in two-man flights) stood.

Moore's biggest new development was the Bell Jetbelt. The benefits of a jetbelt over a rocketbelt seemed obvious – a jet engine is more powerful and more economical to run than a rocket engine. Calculations suggested the jetbelt would be capable of reaching speeds of up to 85 miles per hour and running for up to 25 minutes. Wendell Moore began to develop the new device in 1965, and by 1969 the jetbelt was completed and ready to go. If the rocketbelt resembled a vacuum cleaner, the jetbelt looked more like a washing machine. The huge silver jet engine was mounted on a backpack, along with a parachute for emergency descent.

On 7 April 1969, Robert Courter Jnr made the first untethered jetbelt flight at Fort Myer, Washington. He piloted the device around a 300-foot long course at an altitude of 25 feet, reaching speeds of 30 miles per hour. Further test flights were performed, and all went well. The jetbelt even made a promotional appearance in an Ovaltine advert, with Courter at the controls. Then, suddenly, the entire Bell flying belt project was thrown into disarray.

Wendell Moore died of a sudden heart attack on 29 May 1969, just weeks after watching the test flight of his latest invention. He was 51 years old. With the jetbelt still largely unproven and the rocketbelt already rejected by the US Army, the

flying belt project was abandoned. The Bell Rocketbelt would never fly again.

In 1970, after performing over 1,200 rocketbelt
flights, Bill Suitor decided to leave Bell and take up
employment with the Power Authority back in
Youngstown, New York. But the story of the rocketbelt, and Suitor's involvement in it, was far from
over.

FOUR
THE TYLER ROCKETBELT

The 1968 Barbra Streisand musical *Funny Girl* would not initially seem to have any relevance to a story about rocketbelts. Yet the technology used to capture the movie's most famous shot was created by a true rocketbelt pioneer. The spectacular shot begins with a close up on Streisand belting out *Don't Rain On My Parade*, then sweeps back and up into the air to reveal that she is standing on a tugboat in the middle of New York's Hudson River. The shot was filmed from a helicopter, and was remarkable at the time because of the lack of vibration. The smoothness of the shot was achieved by utilising a vibration-free camera mount designed by an engineer and aerial photographer named Nelson Tyler.

Tyler, from Van Nuys, near Los Angeles, California, built the first version of the camera mount in his garage. He set up a company called Tyler Camera Systems, and worked on movies like *Ice Station Zebra*, *Paint Your Wagon*, and *Catch-22*, winning Oscars in 1964 and 1981 for his technical achievements in moviemaking. But Tyler had another interest – building rocketbelts.

Tyler first became fascinated with rocketbelts after seeing Bill Suitor fly the Bell belt at Disneyland in 1965. Tyler photographed the belt almost obsessively, noted technical details, studied rocket propulsion, and purchased a copy of a key valve from the same manufacturer that built the original Bell Rocketbelt parts. Then, in an incredible feat of engineering, Tyler built his own belt, at a cost of around $15,000, having sold his sports car to fund the project. He tested it in around 70 tethered flights, but continually met problems. Then luck intervened

On the very day in 1970 that he left Bell, Bill Suitor was handed a photograph showing Nelson Tyler and his Tyler Rocketbelt. Suitor contacted Tyler at noon that day, and was thrilled to learn that the new project had been inspired by his own Disneyland flight. He remembered Tyler from the Disneyland stint, as a very persistent gentleman who insisted on having multiple photos taken next to the rocketbelt. What Suitor hadn't realised was that, while posing for the photos, Tyler had been holding a small measuring scale in his hand. Suitor offered to help Tyler complete the project, and by 3pm that afternoon he was on a plane to Los Angeles.

Upon arriving in California, Suitor began to tune-up and test the rocketbelt. After two successful tethered tests the Tyler Rocketbelt was ready to fly. Not that Tyler had any great plans for his creation – he had considered it purely as a hobby. But Clyde Baldschun, the promoter who had worked with Bell, did recognise the commercial possibilities of the Tyler Rocketbelt. He teamed up with Tyler to make entertainment bookings for the new rocketbelt and, over the next decade and a half, the belt made around 120 commercial flights.

Tyler himself performed flights for a couple of television advertisements, but for the most part he relied upon Bill Suitor to fly the belt. It was a match made in heaven, and Suitor was a bona fide rocketman again. Suitor's first major flight of the Tyler Rocketbelt was at the American Football Pro Bowl in 1971. He made scores of subsequent flights, including one in 1982 at the World's Fair in Knoxville, Tennessee. During what he calls his 'oh shit' flight, Suitor unintentionally flew to an unprecedented altitude of 140 feet.

'I buzzed a ski chairlift with tourists in it,' said Suitor. 'They didn't realise I was coming. One woman dropped her camera to the ground, and another woman almost fell out, 40 feet above the sidewalk. Needless to say I was advised not to do that again! But it was fun, especially looking at their faces as I appeared.'

Suitor and Tyler prepared for the Knoxville demonstration with a test flight on wasteland in Los Angeles. On an adjacent lot, construction was beginning on a velodrome in preparation for the 1984 Olympics. The two men surveyed the construction work and Suitor remarked that it would be fitting if he could end his career as a rocketman with a triumphant rocketbelt flight at the Olympics.

In fact, Suitor and Tyler's partnership ended immediately after the Knoxville flight. Suitor found that Clyde Baldschun had misled him over payment and insurance, and he subsequently quit. He returned to his home in Youngstown to take care of his family responsibilities, telling Tyler that his days as a rocketbelt pilot were over.

Tyler had already recruited a second pilot to fly the rocketbelt. Kinnie Gibson was a stuntman, now best known for his work as Chuck Norris's double for the TV show *Walker, Texas Ranger*. Howard 'Kinnie' Gibson, who shared Norris's rugged, bearded look, had competed in motor cross tournaments, skydiving championships, and hot air balloon races, before touring with his idol – ultimate stuntman Evel Knievel. Gibson, from Carrolton, Texas, took up professional stunt work, and became known for his aerial stunts involving light aircraft and wing walking. He began to fly the Tyler Rocketbelt in the early 1980s, and flew at demonstrations around the world. He learned fast, and replaced Suitor as the full-time rocketbelt pilot in 1982, with Suitor's approval. And, two years later, Gibson looked set to pilot the Tyler belt in the biggest rocketbelt demonstration ever.

Clyde Baldschun had been contacted by Tommy Walker, the special effects director of the Olympics opening ceremony. Walker wanted to book a stunning attraction that would seize the attention of the spectators in the Olympic stadium and television viewers worldwide. He wanted to book the Tyler Rocketbelt – but the offer came with certain stipulations. The flight needed to be precisely timed and executed. The rocketbelt must take off and land on cue, with a specific turn, and an accurate landing on an exact spot. Although an excited Kinnie Gibson had told his family about the big flight, and had even prepared his own press release, Tyler and Baldschun weren't sure the rookie pilot could pull this one off. Tyler recalled his conversation near the LA velodrome two years earlier. He called Bill Suitor.

Suitor was reluctant to leave Youngstown. He was busy building an extension on his house, and plastic sheeting was tacked over demolished walls to keep out the elements. He had also recently been promoted in his job at the New York Power Authority, and was unable to get time off work. But Tyler and Baldschun were persistent. Baldschun called in a favour from a friend in authority, and put in a request to the Governor of New York. The Governor arranged for Suitor to be given all the time off he needed. Finally, after Tyler and Baldschun agreed to let him bring his wife along, Suitor accepted the chance to fly the Tyler Rocketbelt one last time. It would be the most famous rocketbelt flight ever made.

On Saturday 28 July 1984, a worldwide audience got its first close-up glimpse of the amazing rocketbelt when Suitor flew in front of 92,000 spectators and several billion television viewers at the LA Coliseum for the opening ceremony of the 1984 Olympic Games. For the first time, the rocketbelt was on the world stage. Dressed in a white jumpsuit with orange stripes and gold stars, and a white helmet bearing gold Olympic rings, Suitor stood on a flight of steps under the Olympic torch at one end of the packed arena. A sign reading 'Welcome to LA 84' was fixed to his chest with safety pins, and another was stuck to the back of the rocketbelt.

The sky was cloudless and bright, and the coliseum was bathed in hot sunshine. Flags ringed the stadium perimeter, and thousands of white and gold balloons lay around the running track waiting to be released. Hundreds of performers crouched on the playing field in preparation for the ceremony. Spectators, in short sleeves, hats and sunglasses, clutched cameras in anticipation, and a cacophony of excited chatter filled the air. Then it began.

'To an audience of two and one half billion that stretches around the world,' began the stadium announcer, 'a warm welcome from the citizens of California!' There was a huge cheer. And then Bill Suitor did his stuff.

'It was, in a word, awesome,' said Suitor. 'As I stood up there 90 feet above the field waiting for my signal to take off, the announcer said, "To an audience of two and one half billion…" Holy shit! He said *billion*! I said to myself: You've done this hundreds of times and never fallen nor failed, so don't go on your ass now in front of all these people!'

'Well, there he is,' proclaimed ABC TV announcer Jim McKay, as Suitor took off over the crowd, 'Jet Man – flying through the stadium, no wires, no tricks, just as you see it. What a beginning!'

Spectators pointed to the sky and watched in awe as Suitor soared overhead. TV viewers sat openmouthed and glued to their sets as he made a graceful turn against a backdrop of camera flashes and waving flags. Then he landed perfectly on his mark on a wooden landing platform, offering a salute to President Reagan up in the stands. The crowd erupted in a hair-raising roar that makes TV footage of the flight particularly memorable to watch. Suitor's wife Cheryl ran over to give her husband a kiss before he was hurriedly escorted from the stadium, allowing the opening ceremony to continue behind him. The flight only lasted 17 seconds, but it was the biggest of Bill Suitor's life.

'It was like a blur,' he said. 'It whizzed past me like a lighting bolt. Then it was all over. But, wow, what an honour.'

The impact of Suitor's flight on the public consciousness was immense. At that moment, everyone everywhere seemed to be talking about the rocketbelt. The *Los Angeles Herald* summed up the reaction in its Sunday edition, printing a photo of Suitor and the belt superimposed over a large one-word front-page headline: *WOW*.

The Olympic opening ceremony committee paid Clyde Baldschun over $7,500 to secure the 17-second flight, equating to $441 per second, or, in effect, an hourly rate of over $1,500,000. Suitor was paid just his standard fee of $1,000, but he considered the honour made it worth it. He may have held a different opinion if he had known in advance that he was uninsured for accidents. Already angered by the Knoxville flight, Suitor had only agreed to perform the Olympic flight on the promise of a sufficient insurance policy. Baldschun had assured Suitor that everything was in order, but in fact it appeared that no adequate insurance policy had been taken.

Baldschun's rocketbelt booking concern had been struggling in the run up to the Olympics, and the belt itself was in a state of disrepair. But the chance to demonstrate the belt in front of potential customers from around the world at the opening ceremony was simply too good to miss. Baldschun reached out to his many contacts and raised around $10,000 in order for the belt to be repaired. The repairs were successful, but it seemed there was no money left over to pay for the pilot's insurance.

When Suitor found out he had been lied to, and had made the flight without insurance, he was furious. He angrily confronted Baldschun, and stated he would never fly the Tyler Rocketbelt ever again, keeping a promise to Cheryl to retire from life as a rocketman. He had, as he had prophesised to Nelson Tyler, ended his career as a rocketman with a triumphant rocketbelt flight at the Olympics. He returned to Youngstown, to his family, his house extension, and his work for the Power Authority. Bill Suitor's career as a rocketman was over, at least for the moment.

Not long after the 1984 Olympics flight, after receiving, it was claimed, around 800 enquiries, Nelson Tyler sold the Tyler Rocketbelt to the Gröna Lund theme park in Stockholm, Sweden. Kinnie Gibson was employed by the park to fly the belt in several demonstrations and, when he returned to the US to fulfil a stunt contract, the park had no use for their pilotless belt. Gibson arranged to procure it from them, and returned home with his new toy. Gibson was now ready to set out on his own as the world's only rocketman.

But there were a couple of problems. Gibson was reluctant to pay high prices for compressed nitrogen cylinders. He used scuba tank compressed air instead, despite its low-purity, with the result that the rocketbelt lost power after around ten seconds. Gibson found this out while filming an advertisement for British Airways in London. He was supposed to fly over the tail of a BA jumbo jet and land safely, but midway through the flight he was forced to make a swift emergency landing, crashing onto his knees on the tarmac runway. A similar accident occurred just a few weeks later during a demonstration flight in Cairo.

Another problem was that the 90-percent H_2O_2 required for the belt was no longer easily available in the US. Gibson was forced to buy 88-percent H_2O_2 from a German company. But the H_2O_2 he purchased had stabilising agents in it, rendering it highly unpredictable as rocketbelt fuel.

Gibson pressed ahead anyway, but his first test flight with the new fuel, in Philadelphia, went horribly wrong. Unbeknown to Gibson, the stabilising agent had ruined the catalyst beds in the rocketbelt engine. Gibson made a stuttering lift-off, but the belt banked and gave out. Gibson fell, and broke both the belt and his knee. He was lucky to survive, and was hospitalised for some time. Gibson sued the German chemical company that supplied the H_2O_2 and won a $250,000 settlement. As a result of the verdict, chemical companies around the world stopped selling H_2O_2 to individuals, a change that would affect all future rocketbelt builders.

Once he had recovered from his injury, Gibson set about repairing the Tyler Rocketbelt, and building an H_2O_2 distillation lab so he could manufacture his own fuel safely. To help him with these complicated tasks he turned to a couple of friends, Brad Barker and Larry Stanley. With hindsight, that was a big mistake. The three men wouldn't be friends for very much longer.

FIVE
THE OIL FIELD INCIDENT

It was early one morning in 1990 when the car arrived at the oil field near Houston, Texas, and hurtled along a dust track past the wellheads and drilling machinery.

Among the passengers in the car was 35-year-old Brad Barker, a former insurance salesman from Illinois. Barker had movie star looks and a charismatic personality. But those who had crossed his path knew that behind his handsome and charming façade was a volatile temper. He was a complicated man, they said, a regular churchgoer liable to outbursts of violence. Barker had been recruited to retrieve equipment from the oil field after a dispute. On his lap lay a baseball bat. It was clear Barker was not there to negotiate.

Bradley Wayne Barker was tanned and in shape, with dark eyes and thick brown hair. He was most often seen in jeans and an unbuttoned polo shirt, his hair carefully swept back. Barker had lost his father as a young boy, but remained close to his mother and brother. He was a wanderer, and rarely stayed for long in any one place or job. But if there was anywhere he called home it was Houston, a gritty and humid big city dominated by oil excavation and space exploration.

Amid the Bayou City's contrasting cityscape of skyscrapers and waterways, black gold drove the economy and the NASA Space Center drew tourism. International finance, computer technology and science research also prospered in the area. A culturally diverse population had been attracted by Houston's reputation for opportunity and innova-tion. The city seemed to be continuously growing and changing, but cowboys still strode along side-walks, and the smell of barbequed steak still drifted reassuringly from restaurant doorways.

Barker had driven to the oil field on behalf of his close friend, rocketbelt pilot Kinnie Gibson. The pair met in 1975, when a 20-year-old suited-up Barker walked into Houston's Central National Bank looking for a job. He was interviewed by Gibson – then just 19, stuck in a desk job, and yet to fulfil his dream of becoming a professional stuntman. The pair clicked immediately. They were almost the same age, shared many interests, and found it easy to get along. Barker aced the interview, and the pair began to work side by side.

Both Barker and Gibson had been drawn to Houston's high-rise business district by a booming job market, but neither was particularly interested in financial work. They were in it for the money, and compensated for spending their weekdays in an office by spending their weekends outdoors – travelling, skydiving and flying planes. Barker bought a small airplane, a Cessna 210 Centurion, and the two men flew it out of the nearby League City airport. They partied together at Houston bars, drank, met girls, and happily enjoyed the best years of their lives. They also went through defining moments together – they acted as best man at each other's weddings, and their sons were born within 24 hours of each other. Barker and Gibson were best friends, and it seemed nothing could change that.

By the beginning of the 1980s, both men had drifted away from selling insurance. Houston's business district was no longer the draw it had once been. Oil prices were beginning to fall, and Houston's much-hyped economy was starting to dwindle. The fourth-largest city in the US was about to experience an exodus in its population.

Gibson was one of those that left. He wanted to be active, not stuck behind a desk. He loved extreme sports and flying, and wanted to make a living from those types of activities. He decided he wanted to be a stuntman. Gibson moved to Los Angeles and set about finding a way into professional stunt work. Then he was offered the chance to fly the Tyler Rocketbelt. Gibson became a rocketman.

Barker stayed in Houston and moved between various jobs, in the oil industry, in auto shops, and in the nightclub business, none of which paid as well or as regularly as his old insurance job. He knew he could probably have followed Gibson into stunt work if only the opportunity had arisen. Instead, he enviously watched the progress of his best friend, and made sure he kept in touch.

In 1981, Gibson invited Barker to watch him fly the rocketbelt at a demonstration near Mexico City. He knew Barker had fallen on lean times since leaving the insurance business, so he offered to pay his airfare. Barker accepted and flew south of the border, where he spent a week with Gibson and the rocketbelt. And he immediately became enchanted with the amazing flying machine.

In fact, Barker had been interested in rocketbelts ever since seeing the opening sequence of *Thunderball* in a local cinema back in 1965. He recalled that as a seminal moment in his life.

'I was nine years old, in a little town in Illinois,' he said. 'My father had been killed in a car accident about eight months before that, and I just remember being a depressed little kid. Then I saw the rocketbelt in the theatre, and I just remember being fascinated with it.'

After the week in Mexico, Barker returned to Houston, and continued to move between jobs. But he never forgot about the rocketbelt. Then his marriage fell apart, and there was no longer anything left for him in Houston. Barker packed a bag and left the Lone Star State for the Golden State.

In California, Barker hooked up with Gibson and was given a job helping to maintain the rocketbelt. Barker was technically proficient, as a result of experience gained in his variety of jobs, and he quickly learnt how the belt was prepared and operated. Gibson had Barker help him repair the belt after the accident in Philadelphia. But for help building a new H_2O_2 distillation lab Gibson turned to another friend, Larry Stanley.

Thomas Laurence Stanley was an aeronautics buff from Sugar Land – a Houston suburb named for its Imperial Sugar factory. He met Barker and Gibson while skydiving at the League City airport in the 1970s. Dark-haired with a chunky moustache, Stanley was an entrepreneur who was always looking for original ways to make money. His family owned an oil field, and the wealth it generated allowed him to dabble in businesses involving his key interests of computers and aeronautics.

Stanley was ten years older than Barker and

eleven years older than Gibson, and never became as close to the pair as they were to each other. Stanley liked to ski but, unlike outdoor types Barker and Gibson, he wasn't particularly athletic. He was primarily a techie, but the three men still became firm friends. Barker trusted Stanley enough to loan him his Cessna plane on numerous occasions. And Gibson invited Stanley to watch him fly the rocketbelt, and also formed a hot-air balloon business with him, referred to by Gibson as one of the biggest and most lucrative in the world. But the balloon business didn't last long. After it deflated, Stanley headed back to his family's oil field.

Then Barker and Stanley fell out. Barker was very fond of his Cessna, a turbo-charged aircraft originally worth as much as $180,000. In 1986, Barker claimed that the plane disappeared while on loan to Stanley. Barker was unable to locate either the Cessna or Stanley, so he contacted the authorities. According to Barker, an FBI Agent called on the telephone the next day and told him, 'Your airplane is probably at the bottom of the Gulf of Mexico with a load of pot on it.'

According to Barker, the Agent referred to Stanley as being 'a known drug trafficker and smuggler' who had been 'investigated by every drug enforcement agency on the planet.' There is no available record of Stanley ever being charged with, or being investigated for, drug offences. Whether the drug claim was true or not, Barker said the FBI's hands were tied regarding the plane.

'They really couldn't do much,' said Barker. 'If I had never let him use the plane it wouldn't have been a problem, but because I had loaned it to him in the past it was basically my word against his.'

As it turned out, Barker's plane wasn't at the bottom of the Gulf of Mexico. It was in a hangar in Seattle, Washington. According to Barker, Stanley had added a long-range fuel tank and made other modifications to the Cessna. The implication from Barker was that the bigger fuel tank had been fitted to allow drug trafficking trips to South America. There was an outstanding bill of $30,000 tied to the plane, the consequence of which was that the cash-strapped Barker couldn't afford to keep it. He was reluctantly forced to sell his prized airplane.

'When the thing was finally sold,' said Barker, 'the money that was brought in was used to pay off the outstanding bill.'

So Barker was left with nothing. He had lost his plane, and he figured Stanley owed him the value of it. But Stanley was nowhere to be found. Almost four years would go by before Barker would see his former friend again.

Gibson and Stanley also fell out. Gibson had agreed to invest several thousand dollars in the Stanley family's oil field in return for 50 percent of the profits. But there was a disagreement, and

Gibson never saw any return. According to Stanley, Gibson failed to meet the terms of a contract the pair had drawn up, and the contract was terminated. Stanley and Gibson began to feud.

In early 1990, Barker was working on the rocketbelt in Los Angeles while Gibson travelled to the Philippines to do some stunt work on a movie. Shortly after Gibson left, Barker received a tearful phone call from Gibson's wife, Sheri. According to Sheri, Stanley had broken into Gibson's storage facility in Houston and stolen some equipment related to the rocketbelt. Sheri said Stanley had told her not to bring anybody out to try to retrieve the equipment, or else, he had said, 'somebody's gonna get hurt.'

Despite Stanley's warning, Barker immediately flew out to Houston and called up a friend, a karate instructor and black belt called Rob Fisher who once worked at a nightclub Barker had managed. Then Barker, Fisher and Sheri Gibson drove out to the oil field. Kinnie Gibson's two brothers followed them out in a second car. Gibson was a good friend who had done a lot for Barker. Stanley was anything but. There was no way Barker was going to let Stanley screw Gibson over.

One of the oil field's hired hands, Bernie Robinson, watched Barker's car approach in a cloud of dust, and stepped forward as it slowed to a halt. Then the doors flew open and Barker and Fisher burst out. Robinson was a former Navy Seal, and he'd been warned by Stanley to be ready for trouble. According to Barker, Robinson swung at Fisher and then tried to make a dash for his car, where he kept a loaded pistol, but, Barker claimed, 'Rob went into his karate routine and, in literally three or four seconds, just beat the shit out of this Navy Seal.'

Then Kinnie Gibson's brothers showed up. According to Robinson, who said he did not throw a punch before being knocked to the ground, three of the men held him down while Barker stood over him with a baseball bat. Then, Robinson claimed, Barker lifted the bat into the air and brought it crashing down on his legs.

'Where's Larry Stanley?' asked Barker.

Robinson could only groan with pain.

Barker whacked him again with the bat. 'Where's Larry Stanley?'

'He isn't here,' said Robinson.

'That's a shame,' said Barker. 'I'm very upset with Larry Stanley, and I'd like to hit him some.'

It seemed Stanley was again going to prove elusive. Then a car pulled up, and out stepped the man himself. Stanley spotted Sheri Gibson, and began to argue with her. He didn't realise she had brought friends. Then Barker stepped into view.

'Stanley,' said Barker, 'I'm gonna ask you one time where Kinnie's equipment is, and, if you don't tell me, you're not going to like what happens.'

Stanley looked at Barker, and the baseball bat, and said, 'I'll take you there.'

Barker followed Stanley to a nearby storage facility and recovered all of the equipment without further incident. Oddly, Barker said he did not mention the Cessna and the related $30,000 debt, although this was the first time the two men had met in the four years since the plane had gone missing. Instead, Barker quietly packed up the equipment and took it back to California.

Later that day, Barker took a phone call from Kinnie Gibson in the Philippines.

'I love you brother,' said Gibson. 'I can't tell you how much I appreciate you getting my stuff back.'

Gibson would not always regard Barker with such fondness.

Still based in Los Angeles, Barker continued to work with Gibson on his rocketbelt. Gibson won a lucrative contract to fly the belt in front of hundreds of thousands of music fans on pop star Michael Jackson's 'Bad' tour. At the end of every performance, Gibson switched places with Jackson and blasted into the air, creating the illusion that the pop star was flying from the stage. Gibson was paid $25,000 plus expenses for each flight, and he made almost $1 million from the full Jackson tour. Gibson also won a contract to fly the rocketbelt at twenty different events at the Disney World theme park in Orlando, Florida.

In the summer of 1990, Gibson was preparing for a Disney World performance when he accidentally damaged the rocketbelt. He was performing his usual checks on the device, but was unaware that a high level of gas pressure had built up in the throttle valve. He unscrewed the cap, and the trapped pressure fired a piston out of the valve at high speed. The piston shot over Gibson's shoulder and lodged in the wall behind him. Gibson was unharmed, although shocked, but the piston was severely damaged.

According to Gibson, the six-inch long throttle valve was the key to the rocketbelt's secret design. There were no spare parts for the Tyler belt, and the valve could not be repaired in Orlando. So Gibson asked Barker to hurry the valve up to Houston, to be machined by specialists there, and return it before the performance the next day.

Barker took a flight to Houston, fully aware that he had in his possession a crucial and valuable piece of the rocketbelt. So crucial and valuable that Barker decided to measure and record the throttle valve's dimensions. Then he had the valve repaired, flew back to Orlando, and returned it to Gibson in time for the performance. The Disney World flight went ahead as planned, and Gibson remained unaware of best friend's keen interest in the valve. If the throttle valve was the key, then Barker had just unlocked the secret of the rocketbelt.

SIX
THE UNLIKELY PARTNERSHIP

Joe Wright left Michigan after graduating from high school in the autumn of 1979, with unemployment having swept through the once dominant industrial state. Like thousands of other displaced souls, Wright, aged just eighteen, travelled from the 'Rust Belt' of the Midwest to the 'Sun Belt' of the South. And, like Brad Barker, Wright ended up in Houston.

Wright, of average height and slight build, was extremely bright, although he had never done particularly well at high school. He was keen to kick-start his life and start up his own business. It was tough for him to leave his friends and family behind, but Wright knew his future lay outside of

Michigan. The area's dwindling economy offered few prospects. Wright believed Houston's economy was still booming. It was a land of opportunity. For Wright, this was the beginning of a wonderful adventure.

The adventure got off to a discouraging start. Wright spent the first few weeks in Houston sleeping in the back of his car. But he had worked in an audio shop while in high school, and that experience got him a job installing car stereos. Wright had a great work ethic. He was an extremely hard worker, had never taken a sick day in his life, and was always the first to volunteer for overtime.

Wright's dedication to the job won him a promotion, and he eventually worked his way up the ranks to manage the audio shop. From 1983 he also ran his own car stereo installation business out of his garage. That business took off, and Wright decided to pursue his dream and set out on his own. He quit his job and prepared to set up his own business.

In 1984, Wright, in partnership with close friend Bryan Galton, set up a shop called Car Audio Plus in a white concrete building at a strip mall on Houston's busy Spears Road, selling and fitting car stereos, auto alarms, and car phones. Wright ploughed all of his savings into the business and devoted huge amounts of energy into making it a success. He was meticulously well organised. His inherent professionalism even extended to having his casual jeans smartly pressed. And his hard work and keen business sense paid dividends. The shop attracted a lot of customers, and soon began to make good money.

When Galton decided to leave Texas for Florida in 1986, Wright bought out his former partner's share in the company and became the sole owner of Car Audio Plus. Within a few years, the business had expanded to take up three quarters of the strip mall. At its peak, Car Audio Plus employed 22 people and had an annual turnover of over two million dollars. Wright, still in his twenties, was a success.

Money was no longer a problem for Wright. He was now a high roller who could buy virtually anything he wanted. He ploughed a lot of his profits back into the business, and still had cash to spare. But he was also well known for his loyalty and generosity. He regularly handed out loans and gifts to his employees. On one occasion, he handed $5,000 to one of his young installers to pay for dental work, and then refused to allow the installer to pay him back. He liked to barter with customers and win their friendship and loyalty.

Then he did a favour for Brad Barker. Barker loved all kinds of electronic gadgets, and visited Car Audio Plus shortly after it opened looking for a top-of-the-line car stereo system for his Jaguar. Barker was still enjoying the high life his insurance job had afforded him, and he had money to spend, although he still liked to bargain. Eager for Barker's custom, Wright cut him a good deal, and installed the system for him. Barker was extremely grateful, and the pair became friends.

Wright found it easy to make friends. He was the kind of person that parties revolved around. He was a funny guy, with a broad laugh that could bring the house down. He loved jokes, and kept a notebook full of them with him at all times. When he heard a good joke, he laughed in his distinctive style, pulled out a pen, and jotted it down for future reference. He was very easy to get on with and he never held grudges. If Wright had a disagreement with someone, the worst they could expect was to receive the silent treatment for a short while, and then it would be forgotten.

By 1990, Joe Wright, still just 29, was an established and successful businessman. It seemed the only way was up. Wright took time out to visit Barker, now working for Kinnie Gibson in Orlando, and was thrilled to get a close up look at the rocketbelt. But Gibson wasn't thrilled when he caught Barker and Wright videotaping the belt. He told them, in no uncertain terms, that the design of the rocketbelt was top secret. Wright was sent home, but he and Barker remained close friends.

A few weeks later, Barker telephoned from Orlando to ask Wright for another favour. Barker was having problems with Gibson. Barker often brought his young son to work, and Gibson complained that the kid disrupted the work. This situation rumbled on for some time, and the animosity between Barker and Gibson grew. The pair argued and their 15-year friendship hit the rocks.

'We had a bit of a falling out,' admitted Barker.

It would not be the last falling out Barker would be involved in. But it was an untidy and uncomfortable end to a friendship, and it was clear Barker could no longer work for Gibson. He quit, and called Wright for help. Wright drove down to Orlando to pick up Barker and his son, and brought them back to Houston.

Then, said Barker, 'I just basically decided to build a rocketbelt.' His motive was clear. 'It was strictly for money, there's no doubt about that,' he said. 'I'd worked with Gibson off and on for years. There were times when he made no money off it, and there were times when he made some pretty good amounts of money. And I'd always been intrigued by the rocketbelt, so it was just kind of a challenge. I wanted to see if I could build it, but I expected to make some money out of it also. So it was a business thing.'

Of course, Barker had nothing else to do. He had been away from Houston for years, but little had changed. There were still few opportunities for him. Gibson's job offer had rescued him from a cash-strapped life that he had no intention of going back to. He didn't want to return to the short-term jobs he had bounced around before getting involved with the rocketbelt. And now he had seen the rocketbelt, become somewhat fixated with it, and watched Kinnie Gibson get rich off the back of it. If Gibson could do that, why couldn't Barker? All he needed to do was to build his own rocketbelt.

Of course, that was easier said than done. Gibson had not built his rocketbelt, but had acquired it from Nelson Tyler. Aside from the pioneering engineers at Bell, the brilliant Tyler had been the only man ever to successfully build a working rocketbelt. Nevertheless, seemingly undaunted by the task at hand, Barker set to work almost immediately, using the information and contacts he had obtained while working on the Tyler Rocketbelt.

'I'd been working with Kinnie,' he said, 'so I knew 99.9 percent of everything about the belt, but there were a few things I didn't.'

What he needed were a couple of partners.

First, Barker approached Joe Wright. Barker knew Wright had a keen business brain and was skilled at marketing, and he also had useful contacts with various technicians and engineers, including some at NASA. He would be useful to the project. He was also a close friend, and Barker was running out of those. But, despite Wright's enthusiasm and skills, Barker knew the two of them alone could not successfully build a rocketbelt.

Barker needed someone who could offer heavy financial backing. So, in early 1991, Barker began to call Larry Stanley. This was despite the angry confrontation at the oil field, and Barker's claim that Stanley had stolen his plane. Barker and Stanley were most definitely no longer friends. But Barker was willing to overlook that fact in the hope of getting his precious rocketbelt built.

'The guy owed me,' said Barker. 'He owed me for the plane. I'd lost close to $30,000 on that deal. Basically I contacted him to see if there was any way I could get my money back.'

Stanley must have been surprised to receive a call from a man who had threatened him with a baseball bat on their last meeting. But Barker quickly attempted to make peace. He told Stanley that the incident at the oil field had all been down to Kinnie Gibson. Gibson had ordered him to go to the oil field and retrieve his equipment by any means necessary, he said. Barker explained that he too had fallen out with Gibson, and was looking to set out on his own.

Then Barker told Stanley he was planning to build a rocketbelt. Barker explained that he had the knowledge to produce the key rocketbelt throttle valve, and also had a portfolio of high-resolution photographs showing all of the external design features that he had obtained from an acquaintance of Gibson. Importantly, Barker also detailed the huge earnings Gibson had made from his rocketbelt. At this point, Stanley's curiosity was undoubtedly aroused.

Both Barker and Stanley had become very interested in rocketbelts through their association with Gibson. They were both intrigued by the challenge of embarking upon a quest to build such an iconic device. And Barker knew that Stanley felt he had been wronged by Gibson and wanted revenge. But the key motive was money. In partnership, Barker told Stanley, the pair could build a new rocketbelt, take over some of the lucrative contracts, become very rich, and put one over on Gibson in the process.

Barker knew Stanley could borrow money from his family to invest in the project. Barker estimated that the belt would cost around $200,000 to build. Stanley lived in a large house in Sugar Land with his wife and her family. He drove an expensive Toyota SUV and wanted for little because of the family's income from the oil field.

Stanley certainly seemed to be interested. As a challenge and a moneymaking opportunity it seemed like something he would like to do. And the project offered a chance to get even with Kinnie Gibson. But nagging doubts troubled him. After all, just months before, Barker had attacked one of Stanley's employees with a baseball bat.

'I had some strong reservations,' Stanley said in

an affidavit he would later present in court. 'I was very unsure if I should embark on a new venture with a person such as Barker, who seemed to have a violent nature.'

Barker, too, had severe doubts over working with Stanley after the pair's chequered history. And he would come to regret his decision to ignore those doubts. He was proposing to partner up with a man he had accused of stealing his plane and using it to traffic drugs. The two men were now more enemies than friends as the result of a long and complicated past.

'We go way back,' Barker said. 'I'm about half an idiot for even doing business with the guy after all that happened, but for that I take full responsibility.'

Despite mutual doubts, the appeal of the rocketbelt, the lure of the dollars, and the chance to put one over on Kinnie Gibson were obviously too much for either man to ignore. After several phone calls were exchanged, Barker arrived at Stanley's Sugar Land home to discuss the project in person. The negotiations were surprisingly straightforward. Stanley offered that, if Barker let him in on the rocketbelt project, he would pay back the money he owed him for the plane. Barker agreed. The men were now partners in their ambitious quest to build a rocketbelt. It was a partnership built on greed and revenge. It couldn't last.

SEVEN
THE ROCKETBELT 2000

The American Rocketbelt Corporation (ARB) was founded in March 1992, with Brad Barker and Larry Stanley going into the project as 50/50 partners. The pair planned to make an improved rocketbelt called the Rocketbelt 2000, or RB-2000. New techniques and materials were available, and they were confident they could improve upon the old Bell and Tyler belts. Those belts had been built on relatively low budgets with surplus parts. Barker and Stanley aimed to make the RB-2000 lighter and load it with more propellant so it could fly longer and further. The original Bell Rocketbelt had only flown for a maximum of 21 seconds per flight. They were certain they could improve upon that.

Despite their confidence, the task Barker and Stanley were undertaking was immense. After numerous attempts spanning over fifty years, only Wendell Moore's Bell team and Nelson Tyler had successfully built a working rocketbelt. All other aspiring rocketbelt builders had failed. Neither Barker nor Stanley had the technical or engineering know-how of Moore or Tyler. The odds were stacked greatly against them. But that didn't dampen their enthusiasm, and work on the demanding project began right away.

Stanley began to produce technical drawings for the rocketbelt, and for a fuel distillation lab. Then Barker took the drawings to machine shops to have the parts made. Barker stored the parts in his North Houston apartment, taking great care to ensure the security and safety of these precious pieces of the rocketbelt jigsaw puzzle. Soon the apartment was full of valves, frames and tubes, made out of aluminium, stainless steel and high-tensile plastic.

Barker and Stanley relied heavily upon the photos and measurements taken from Kinnie Gibson's Tyler belt, and also consulted engineers, particularly Doug Malewicki of California. Malewicki was a respected aeronautical engineer and inventor, known for his appearances on the US TV show *Junkyard Wars*. He provided engineering analysis and helped design the rocket motor, producing preliminary drawings for the project.

'Brad Barker called me about some design and consulting work,' said Malewicki, who was clearly dubious about the project from the beginning. 'I had him sign a complete legal release before I would start.'

Work continued on the RB-2000 throughout 1992 and into 1993. In addition to seeking engineering advice, Barker and Stanley hired mechanics to help them with the technicalities of piecing the device together. They also had mechanics build a support stand for the rocketbelt, and an aluminium shipping container to store it in. Barker and Stanley would need the shipping container when they began to demonstrate the belt around the world. They were also being security conscious. The safe-like container would ensure the RB-2000 could not be stolen. At least, not by anyone who did not have a key.

As work progressed, the various assembled parts began to take the form of a rocketbelt. Now the partners had something tangible to show for their considerable efforts, Barker and Stanley's initial moneymaking and revenge motives began to be replaced by a genuine desire to complete the device and see it fly. They were beginning to become obsessed with the amazing contraption that was taking shape in front of them.

Joe Wright was around the project from the start, and he began to share Barker and Stanley's enthusiasm and passion for the rocketbelt. In December 1993, Wright offered Barker and Stanley an office and workshop space at Car Audio Plus. Wright was obviously confident the rocketbelt would take off. He offered to defer rent until the belt began to make money. Then he would be paid $500 plus expenses for each month of occupancy. There was no longer room for the rocketbelt and its associated parts in Barker's apartment, and the partners needed more space to build a fuel lab. Barker and Stanley checked out the office and workshop and agreed to accept the offer. They moved into the office, and set up a construction area at the far end of Wright's workshop, out of the way of the car audio business. Wright now became actively involved in the construction of the RB-2000.

The three men then set about putting the finishing touches to the rocketbelt, and also built the fuel lab. But the work did not progress without incident. Stanley was warned by one machine shop owner to keep an eye on Barker, who he said seemed to be an unsavoury character. The shop owner didn't elaborate, but the warning accentuated doubts Stanley already had about his partner. And Wright, such a laid-back guy, found himself subjected to verbal outbursts from Barker over the most trivial of matters. According to Car Audio Plus employees, Wright and Barker could be casually talking about everyday things when Barker would suddenly snap into an unseemly rage.

In June 1994, Barker and Stanley held a meeting at Barker's apartment with Stanley's lawyer brother, Jerry. The project accounts showed the pair had injected almost $190,000 into ARB, nearly all of which had been borrowed from their mothers. They decided, as the rocketbelt project was progressing successfully, to issue equal amounts of stock in ARB – 500 shares to Barker and 500 shares to Stanley. They then immediately held the company's first stockholders' meeting. Barker was elected President, and Stanley was named Vice President, Secretary and Treasurer. They also agreed to appoint Joe Wright as Marketing Director. Then Stanley issued share certificates to himself and Barker. Now the rocketbelt was officially owned, not by Barker and Stanley as individuals, but by the American Rocketbelt Corporation.

The rest of the summer was spent completing the rocketbelt. Finally, in October 1994, after more than two and a half years of hard work, the RB-2000 was finished. And it looked fantastic. The frame and handlebars were bright red, and the fuel tanks and exhaust nozzles were shiny silver. Every part gleamed. Barker, Stanley and Wright had worked long and hard to create this amazing device. But would it work? It was time to put the rocketbelt to the test.

After a careful fuelling process, the belt was fitted to its stand in a cleared space at the back of the Car Audio Plus workshop. Stanley and Wright were both in attendance, but it was Barker who performed the test. Barker stood behind the rocketbelt wearing a protective jumpsuit and helmet and prepared to twist the throttle.

'Here goes,' said Barker.

'Go on, buddy!' Stanley and Wright called, as they retreated to a safe distance.

Then Barker gently twisted the throttle. Ear-splitting jets of steam roared from the nozzles, and the rocketbelt strained against the stand.

'Yeah!' shouted a grinning Barker, over the piercing scream of pressurised gas. 'It's working! I think we know it's not gonna blow!'

Barker allowed the throttle a longer blast, and offered a big smile to Stanley and Wright. Then Barker released the throttle and let the rocketbelt fire for several more noisy seconds until the fuel tanks emptied. The initial test was a success. Now it was time for the next stage of the testing process. The Rocketbelt 2000 made a lot of noise, but would it fly?

Barker and Stanley needed to find a pilot before they could properly test their rocketbelt. Both were keen to fly the belt themselves, once they were sure it would work, but neither was reckless enough to test fly the unproven device. They needed to call upon an experienced professional to test it and train them. And there was only one man for the job.

Bill Suitor had not flown a rocketbelt for over ten years. He was happily living in Youngstown, still working for the Power Authority, and painting and carving wildlife in his spare time. A television company was planning to shoot some footage of the RB-2000, and it was they who first contacted Suitor asking him to test the belt. But Suitor wasn't keen on flying the belt for television cameras. He was no longer willing to put his life at risk for the purpose of entertainment. Then he received a telephone call from Barker.

'I remember he didn't really seem interested,' said Barker. 'I think he thought I was just some kind of idiot. I told him I'd built this rocketbelt and would like for him to come down and fly it. He said he had better things to do.'

But Barker was persistent, and he asked Suitor to take a look at video footage of the initial test. Barker Fed-Exed the video to Youngstown that day. Suitor called back within 24 hours. According to Barker, he was 'really, really impressed'. The tape had clearly aroused Suitor's interest, and perhaps rekindled his great love of rocketbelts. Suitor agreed to travel to Houston to take a look at the RB-2000. Barker picked him up at the airport and drove him to Car Audio Plus. Then Barker led Suitor in to the office where the rocketbelt was fitted to its stand.

'When I opened the door, and the rocketbelt was

sitting there, his mouth just literally dropped,' said Barker. 'He probably did not say a word for about five minutes. He just walked around and touched it, getting fingerprints all over it.'

Barker wore gloves to handle his precious belt, polished it regularly, and was wary of other people touching it. Indeed, he hung a specially made sign on the device warning, 'DO NOT TOUCH THE ROCKETBELT.' Of course this was primarily for safety reasons – adjustments to the belt's settings could do a lot more damage than fingerprints.

After he had finished messing up Barker's precious polish work, Suitor said, 'Do you mind if I name the rocketbelt?'

'No,' said Barker, 'I'd be honoured if you did.'

Suitor studied the belt. The frame was bright red and the fuel tanks were gleaming silver. It looked very impressive. Then he said, 'I'd like to call it Pretty Bird.'

'I remember thinking, God, please don't call my rocketbelt Pretty Bird!' said Barker. 'But that's the name he gave it.'

As much as Suitor liked the look of the rocketbelt, he thought it could use some improvement.

'The RB-2000 was a beast,' said Suitor. 'It was, like me by 1994, overweight and out of shape!'

Barker and Stanley reckoned that if the RB-2000 carried more fuel it would fly longer, but that wasn't necessarily the case. More fuel meant more weight.

According to Suitor, weight, not fuel, was the key factor in determining how long and how far the belt could fly. Suitor acknowledged Doug Malewicki had improved the motor design, but dismissed the RB-2000's other so-called improvements. Barker claimed the RB-2000 would run for 28 seconds – over 30 percent longer than the Bell and Tyler belts – but Suitor wasn't convinced. In any case, even with that predicted improvement, the flying time of the RB-2000 remained inadequate for any practical purpose.

'All the hype spewed by Barker was just that – hype,' he said. 'Barker was a showman, not a scientist. Had we fooled around with the rocketbelt it could very well have had a run time of 28 seconds, but what have you have really gained? You have just spent a small fortune reinventing the wheel. You still have a machine that is useless for anything other than shows.' So why call it Pretty Bird? 'If nothing else,' said Suitor, 'it was very pretty.'

Although with hindsight he claimed to be largely unimpressed, Suitor was clearly intrigued by the possibilities surrounding the 'Pretty Bird'. He saw the RB-2000 as a challenge, and a way for him to get back into rocketbelts. He studied the belt and made some suggestions, and the dimensions of the belt were slightly redesigned to fit him. Then Suitor agreed to test the RB-2000 and, once he had become familiar with it, train Barker and Stanley to fly it.

Barker and Stanley had both already bought helmets and flight suits, made of temperature resistant material, in preparation.

The next day, Barker fuelled up the rocketbelt and helped Suitor strap it on. The plan was to gently test the thrust capabilities just to the extent of lifting Suitor up onto the balls of his feet. There was great anticipation as Suitor twisted the throttle, and the belt duly produced its thunderous jets of stream. But nothing much else happened. The downward thrust produced just wasn't powerful enough to generate any lift. It was a deflating anticlimax.

'We had some problems,' said Barker. 'It wasn't putting out enough thrust. The throttle valve was flooding the motor.'

The throttle valve was the vital part that was the key to building a working rocketbelt. And it didn't work. This was devastating news. The possession of the throttle valve secret was the crucial factor that Barker had used to persuade Stanley to help him build the rocketbelt. They had come so far, but this major setback threatened to wreck the entire project. It seemed they had wasted many thousands of dollars and countless man-hours on a 'pretty bird' that was never going to fly.

But Bill Suitor was prepared for such an eventuality, and he had a solution. He went to his suitcase and produced a series of technical drawings he had obtained while working for Bell Aerosystems. Among them were detailed drawings of the original throttle valve from the Bell Rocketbelt. Barker took the drawings to his manufacturing contacts and had three new valves machined. Suitor had provided Barker, Stanley and Wright with the final piece of the rocketbelt puzzle.

Once the throttle valve had been replaced, the RB-2000 team prepared to make outdoor tethered flights, with the belt fixed to an overhead cable and held down by safety ropes. This was the first attempt to have the RB-2000 lift its pilot clear off the ground. It was the most dangerous moment during testing so far. If Suitor was nervous he didn't show it as he was fitted into the belt wearing a flight suit and a helmet, in bright sunlight behind the Car Audio Plus workshop.

'If it feels good when it lifts off,' Suitor told Barker and the others, 'I'm going to move forward a little bit, move backwards, and turn around.'

Suitor gave the throttle a quick test burst, and then fired up the belt for lift-off. It started perfectly. A powerful jet of steam shot down from the exhaust nozzles, stirring up a cloud of dust. Suitor lifted gently into the air, hovering a few inches from the ground. Barker and Stanley held onto safety ropes restricting Suitor's altitude. Gradually, they released the tension on the ropes, allowing the rocketbelt to rise higher and manoeuvre within the confines of the tether. Suitor moved the belt forward for a few feet, then turned, headed back, and gently landed on the ground. The flight lasted 20 seconds. Then Barker hurried over to the smiling Suitor and shook his hand. A whooping Wright followed to congratulate the pilot.

'Brad,' said Suitor, 'this thing has got power up the ass!'

The tethered flight was a success. Amazingly, it seemed the RB-2000 would really fly. Barker, Stanley and Wright continued their work on the belt with new vigour, and Suitor headed home.

According to Barker, 'He flew home, we finished the belt, and from then on there was murder, kidnapping, and all kinds of other stuff.'

EIGHT
THE HAMMER ATTACK

The RB-2000 team's dream of flying through the air with a rocketbelt was well on the way to being realised. But the dream began to sour after Cal Jacobs, an engineer who had built brackets to support the rocketbelt motor, came to see Larry Stanley. According to Stanley, Jacobs told him that Brad Barker had been attempting to pad out the bills for the machine work on the parts. Jacobs claimed Barker had asked him to sign for increased charges on the motor support brackets.

When Bill Suitor headed back to Youngstown, and testing was temporarily halted, Stanley examined the company accounts and began to investigate Barker's expenditure, concluding that Barker had padded other bills and faked expenses in order to pocket the difference. Stanley reckoned Barker had defrauded the American Rocketbelt Corporation out of around $30,000.

Looking back on this claim, Barker denied all accusations of cost padding. According to Barker, Stanley had bounced several cheques with parts suppliers. This had embarrassed Barker, who offered to pay one supplier double the outstanding debt. That was all there was to it, and Stanley had no evidence to support his claim.

'There's no truth in it whatsoever,' Barker said. 'It's a joke, an absolute joke.'

Stanley didn't think it was a very funny joke, but he decided not to confront Barker with his accusations. Bill Suitor was due to return to Houston in November 1994 to begin test flying, and Stanley didn't want to disrupt that. What had started as an exciting challenge and a moneymaking opportunity had become much more than that. This was no longer an exercise in putting one over on Kinnie Gibson, or even just a way to make money. Barker and Stanley had in their possession a working rocketbelt – an amazing device that could offer the pair the chance to fly through the air without wings. And that was something Stanley did not intend to jeopardise.

But then Barker and Stanley couldn't agree over who should be allowed to fly the rocketbelt. Stanley had lost some weight specifically so he would be light enough to fly the belt, but Barker still considered him to be too heavy. In any case, neither man was prepared to allow the other to be the first to fly the belt. Both had become obsessed with the device over the long quest to build it, and they considered it far too precious to share.

The disagreement over flying rights began to cause arguments and, according to Stanley, one afternoon in October 1994, the pair began to shout at each other in the Car Audio Plus office.

'You're not flying this belt, lardass!' yelled Barker. 'No one is flying it but me!'

'I have exactly the same right to fly the rocketbelt as you do!' Stanley retorted.

Suddenly, according to Stanley, Barker snapped. He grabbed Joe Wright's 9mm semi-automatic pistol from the office desk and pushed the muzzle into Stanley's forehead.

'Motherfucker!' Barker screamed. 'I ought to kill you right now!'

The pair stood in silence for a few moments. The gun remained pressed to Stanley's head. All that could be heard was the sound of Car Audio Plus employees busying themselves with work on the other side of the office door. It seemed all of Stanley's doubts about Barker's violent behaviour were being been realised. Stanley feared his decision to go into partnership with Barker was about to get him killed.

Then, as suddenly as it had arrived, Barker's red mist cleared. He removed the gun from Stanley's forehead and placed it back on the desk.

'You'd better just get the fuck out of here,' said Barker.

Stanley didn't move, but remained standing against the desk. Barker moved over to the rocketbelt and began to tinker with it. This was the first major argument between the pair in the three years since they had begun to build the rocketbelt. It wouldn't be the last.

Stanley watched Barker in silence for around ten minutes, and then left.

An hour later, Barker came to Stanley and apologised. 'I don't know what came over me,' Barker said.

Stanley attributed the incident to Barker's unpredictable mood swings. 'I understood he was prone to fits of anger with little provocation,' said Stanley. 'Throughout 1994 Barker used Prozac and Zoloft he bought illegally to control his moods, which varied unnaturally from extreme happiness and ebullience to deep despair and moodiness. He treated his moodiness with alcohol and drugs.'

Barker vehemently denied that was the case, and he completely refuted the entire gun incident. 'I swear to you that just absolutely never happened,' he said. 'He said that I took a pistol and pointed it right between his eyes, and he said he didn't react at all. If somebody takes a gun and puts it to my head I'm going to cry like a little girl! Stanley is just... he's an interesting character.'

A few days later, on the morning of 12 November 1994, Barker arrived at Car Audio Plus to collect Stanley's share of a payment to purchase a rocketbelt part. He found Stanley in the office and got right down to business.

'Do you have the money I asked you for last night?' asked Barker.

'No,' said Stanley. 'I can't get the money until Monday.'

'Then write me a cheque,' said Barker.

'I can't write you a cheque without having the money in the bank,' said Stanley. Wrong answer.

'That's not going to cut it,' said Barker. 'I've had just about enough of you!'

According to Stanley, Barker slammed the office door shut, and started towards him. Again, the Car Audio Plus office housed a chaotic confrontation. Barker began to swing a succession of punches at his business partner. Stanley blocked the blows, and grabbed Barker in a bear hug, sending paperwork and office furniture flying all over the room. Barker desperately tried to free himself, then charged forward, slamming Stanley into an internal door, and smashing it open. The pair crashed through the splintered door and landed on the floor in an adjacent office. Barker then jumped on top of Stanley.

'I got hold of Stanley,' said Barker. 'I had him on the ground in a chokehold, and once he quit biting I let go of him.'

Barker pulled himself to his feet, and quickly realised he was injured. A large splinter had split from the smashed door and cut his left hand. Blood was pouring from a large gash. He picked up the telephone and called Joe Wright.

'Joe,' said Barker, 'I've hurt my hand pretty bad. I need you to come to the shop and get me to a doctor.'

Wright arrived a few minutes later, and he took Barker out of the shop and drove him to the local emergency room. Barker's hand was X-rayed, and it was found he had broken a finger. His hand was placed in a splint, and his arm put in a sling.

Meanwhile, Larry Stanley went out for lunch with his son, and then visited Cal Jacobs, the machinist who had first suggested to Stanley that Barker might be padding costs. Stanley told Jacobs he was going to confront Barker. He asked Jacobs, a big man who few would choose to mess with, to come with him for support. Jacobs agreed, so Stanley and Jacobs returned to Car Audio Plus.

It was now two in the afternoon. Neither Barker nor Wright were at the shop, so Stanley and Jacobs waited in the office. When Barker and Wright failed to show, Stanley sat down at the office desk and called Wright at home.

'Joe,' said Stanley. 'I've got something to tell you: Brad is stealing from the company.'

Stanley began to reel out his evidence against Barker for Wright, and recounted his visits to the machine shops. He was unaware that Barker had arrived at the audio shop and was listening in from behind the office door.

'All I can tell you,' Stanley told Wright, 'is that Barker is gonna fuck you the same way he fucked me. But me and Cal are gonna teach him a lesson he won't forget.'

Barker had demonstrated just hours earlier that it was very unwise to make him angry. As he listened to the phone call he became enraged. He stormed out to the workshop, grabbed a toolbox, and pulled out a bright orange lead-filled dead-blow hammer. He cast an unusual figure, his right hand brandishing the hammer, his left arm in a sling and his left hand in a splint, but his furious intentions were clear.

Then Barker charged into the office, and yelled, 'Get the fuck off my phone!'

Stanley stood up and said, 'Fuck you!'

Then, according to Stanley, Barker screamed, 'I'm going to kill you, motherfucker!' and thumped him on the back of the head with the hammer.

Stanley fell, knocking his office chair flying. Barker grabbed him, and the pair began to wrestle furiously. Cal Jacobs leapt to his feet, and desperately tried to separate them, eventually managing to grab Barker in a bear hug. But Barker still managed to raise the hammer above his head, and swung again at Stanley, trapping his hand against his skull with a ferocious blow. Stanley's right ring finger was severed at the knuckle and dangled by a sliver of flesh. Jacobs desperately tried to wrench the hammer from Barker's hand, but failed. He still held Barker tight in the bear hug, and managed to lift his feet clear off the ground. But Barker swung the hammer again, and struck Jacobs on the knee-cap. Jacobs loosened his grip and fell to the ground. Then Barker kicked him in the head.

According to Stanley, Barker continued to hit him until Jacobs eventually managed to grab the hammer and push the pair against the office outer door. Barker was temporarily subdued, and there was silence for a few moments. Then Barker began to yell for help.

Within seconds, a worried Car Audio Plus employee began to attempt to kick the office door down. Stanley, squashed against the inside of the door, pushed backwards. Then the door burst open, smashing off its hinges, and Stanley, Barker, and Jacobs were flung back onto the floor. Jacobs lost his grip on the hammer, and Barker took it back and raised it above his head.

'Jacobs,' Barker said, 'if you ever get into my business again, I will kill you!'

Barker, still wearing his sling and brandishing the hammer, pulled Jacobs to his feet and pushed him out of the office and into the workshop. He ordered Jacobs to go back to his workplace, and asked one of the shop employees to drive him. Jacobs didn't argue.

Then Barker walked back into the office. And Stanley was waiting for him. Stanley grabbed Barker in a chokehold from behind and punched him in the face. Barker swivelled around and punched Stanley several times with his uninjured right hand.

'The last time I hit him he fell over backwards, laying on the ground like a turtle,' said Barker. 'I stood over him and told him to stay on the ground. And he did.'

Stanley's head and face were covered in blood, and his white shirt was soaked red. According to Stanley, Barker had hit him eight to ten times with the hammer.

'Barker almost succeeded in killing me,' said Stanley, 'and had Jacobs not been there to stop him, I am sure he would have killed me before he stopped.'

Barker later said he only hit Stanley twice with the hammer, and only then because Stanley 'got in my face'.

However many times he had been hit, Stanley appeared to be very badly injured, but he still had

Bill Suitor wears the Tyler Rocketbelt at the LA
Olympics, 1984 (courtesy Bill Suitor, photo Liz Tyler)

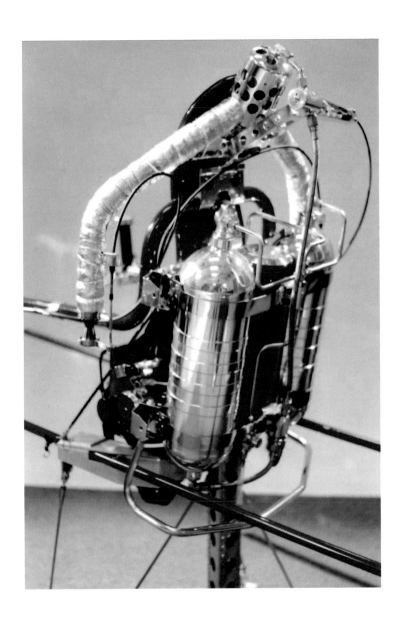

The Rocketbelt 2000 (RB-2000)

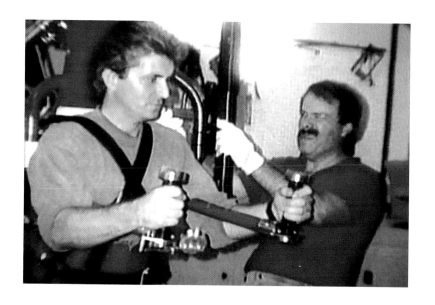

Brad Barker (left) and Larry Stanley
work on the Rocketbelt 2000, 1994

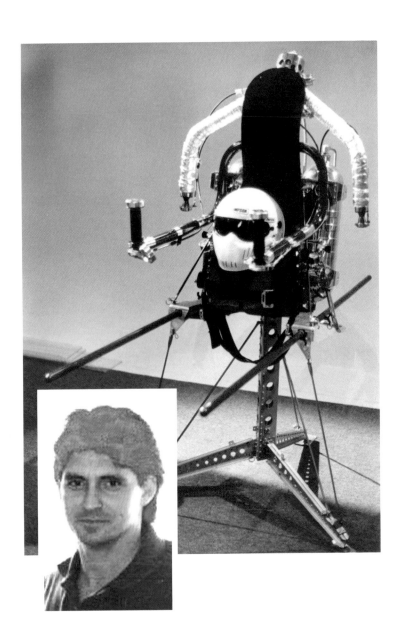

The Rocketbelt 2000, Brad Barker as
shown in Stanley's 1999 affidavit

Joe Wright wears the RB-2000,
Joe Wright, Car Audio Plus

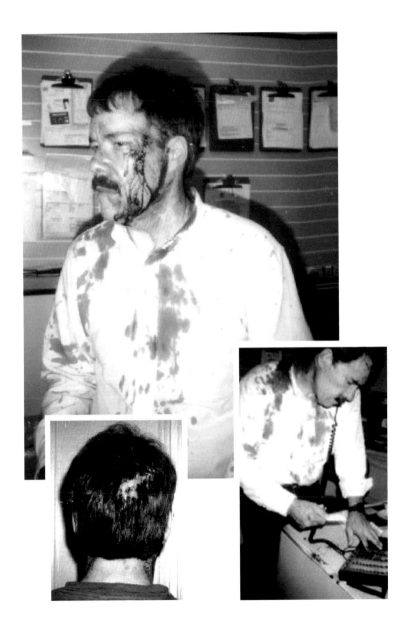

Larry Stanley after the hammer attack, 1994,
Stanley calls 911, displays his head injuries

Bill Suitor tests the RB-2000, 1995

Bill Suitor test-flies the Rocketbelt 2000, 1995

Brad Barker in Houston, 2004 (photo Danny Carr)

Stuart Ross performs a successful tethered flight, and wears his home-made rocketbelt (courtesy Stuart Ross)

Peter Gijsberts and Kathleen Lennon Clough at the 2006
Rocketbelt Convention with original Bell Rocketbelts,
and with Bill Suitor (courtesy Kathleen Lennon Clough)

Pilots and organisers at the 2006 Rocketbelt convention: (L-R) Eric Scott, Nelson Tyler, John Spencer, Peter Gijsberts, Kathleen Lennon Clough, Bill Suitor, Peter Kedzierski, Hal Graham (courtesy Peter Gijsberts)

Eric Scott flies the GoFast! Rocketbelt at the Rocketbelt Convention, 2006 (courtesy Kathleen Lennon Clough)

Eric Scott flies the GoFast! Rocketbelt
(courtesy Kathleen Lennon Clough)

'World's First Rocketwoman' Isabel Lozano tests the
TAM Rocketbelt (courtesy Juan Manuel Lozano)

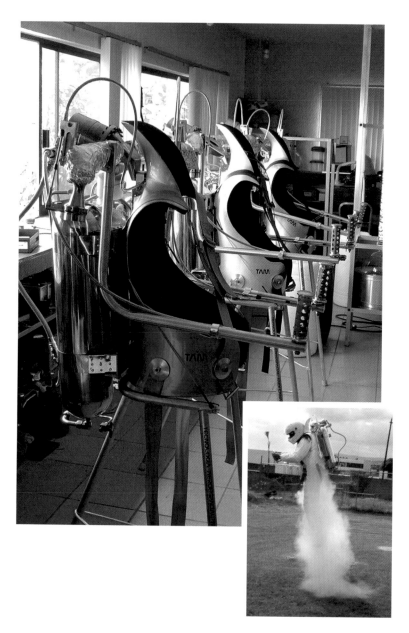

The TAM Rocketbelt in mass production, and testing
(courtesy Juan Manuel Lozano)

the presence of mind to pose for a remarkable series of photographs detailing his injuries. One of the photos showed Stanley dialling 911, with blood streaming from his head and down his neck. He asked the telephone operator for an ambulance and for the Sheriff's Department, and said he wanted to report an assault. But a Car Audio Plus employee, seemingly siding with Barker, grabbed the phone and told the operator to ignore Stanley.

Nevertheless, 25 minutes later, two Harris County Sheriff's Deputies arrived at the scene. Joe Wright had also arrived, and he and Barker met the Deputies in front of the shop. Barker told the Deputies that Stanley had started the fight, but had come off worse. Then an ambulance arrived. The paramedics were shown to the office where they took one look at Stanley and said they needed to get him to hospital immediately. Stanley refused medical attention. Then the Sheriff's Deputies entered the office and told Stanley he was going to be arrested for assault.

'I was flabbergasted that they made this decision without listening to my side of the story,' said Stanley. 'Barker had no scratch on him, while I was standing there with my shirt soaked from shoulders to waist with blood.'

Stanley told the Deputies he wanted to press charges against Barker for felony assault with a deadly weapon. Then Joe Wright attempted to mediate between Barker and Stanley. He told Stanley that Barker would not press charges against him if he agreed not to press charges against Barker. Stanley refused. He signed a refusal of treatment release for the paramedics, and the ambulance left, leaving Stanley dripping blood all over Wright's shop. Then the Sheriff's Deputies arrested both Stanley and Barker, and the pair were taken straight to the Harris County Jail. Both were charged with misdemeanour assault.

Stanley soon began to regret refusing medical help. After loudly complaining, he was finally taken to the emergency room at Lyndon B Johnson General Hospital. By Stanley's account, over a six-hour period he received stitches to the back of his head, underwent a CAT scan, and had his severed finger reattached by an orthopaedic surgeon. Then Larry Stanley was returned to the County Jail – and placed in a cell with Brad Barker.

'I have never understood what purpose the jailer had in mind here,' said Stanley, 'but it was moronic.'

The former friends sat next to each other, Barker with a broken finger and his arm in a sling, and Stanley with a severed finger and bandages across his head. After an uncomfortable few hours, Barker was released at 2.30am, and then Stanley was released at 4.30am. Stanley said he spent the next week flat out on his bed unable to move as he recovered from his injuries.

Perhaps understandably, Barker said the hammer attack was 'pretty much what ended the partnership.'

Two days later, on 14 November 1994, Brad Barker and Joe Wright met with a lawyer friend of Wright, Jonette Anderson. They drew up a lien (the right to hold property until a debt is paid) against the American Rocketbelt Corporation, and Larry Stanley, to the value of $12,000 relating to unpaid workshop rent. Wright's offer to defer rent payments had never been confirmed in a written agreement, and Barker and Wright were now claiming the offer was invalid.

Anderson also set up a new company on behalf of Barker and Wright called Duratron Incorporated, and began to make attempts to transfer ARB's assets – including the rocketbelt – to Duratron. Within days, Anderson announced that Duratron had purchased all of ARB's assets including the rocketbelt, in a public auction for just $10,000.

Shortly afterwards, Brad Barker drove to Car Audio Plus and removed the rocketbelt, the fuel laboratory and all associated equipment. Barker also removed a .357 calibre Winchester rifle.

Stanley was understandably furious. He pursued the rocketbelt through his attorneys, and filed a civil lawsuit on behalf of ARB against Barker, Wright, Anderson, and Duratron for conspiracy and grand theft to defraud through the filing of a false lien. He also pushed the District Attorney to upgrade the assault charge against Barker relating to the hammer attack from misdemeanour assault to first-degree assault with intent to murder. But the most important thing to Larry Stanley was the RB-2000. Stanley had been separated from his precious rocketbelt, and he was willing to go to extraordinary lengths to get it back.

NINE
THE TEST FLIGHT

With Larry Stanley temporarily out of the picture, Brad Barker and Joe Wright continued to test the rocketbelt with Bill Suitor. The first untethered RB-2000 test flight took place behind Wright's Car Audio Plus building on 21 January 1995.

Tethered tests had shown that the rocketbelt could fly and be manoeuvred, but there was no way of knowing how reliable the belt would be when free of the safety ropes and harness. Tethered flights had only allowed the belt to reach an altitude of a couple of feet. From that height, a fall was unlikely to do serious damage to the pilot or the rocketbelt, but this free flight would see the rocketbelt fly at an altitude of over ten feet. A fall from that height could be fatal. There was no margin for error. Nothing could go wrong.

Suitor, wearing a white jumpsuit and helmet, just as he had done at the Olympics in 1984, walked into position with the rocketbelt on his back. The bright sun cast long shadows against the whitewashed walls of the audio shop. A car alarm rang out in the distance. Other than that, there was silence. Call it nervous anticipation, but Barker and Wright were uncommonly quiet as they watched from a safe distance, the fate of their precious rocketbelt in the hands of a man in a jumpsuit.

Suitor briefly released the throttle, creating a small amount of thrust that bounced him up onto his tiptoes. There was a pause. Then he released the throttle again, slightly longer this time, allowing him to lift clear of the ground by several inches, pirouette for a few seconds, and land. It was a noisy, thrilling sight. So far so good.

A second attempt saw Suitor rise six feet from the ground, and manoeuvre in a small circle for a good ten seconds. After refuelling, Suitor twisted the throttle, lifted into the air, and moved in a larger circle. Then the belt was refuelled again, for a last tentative test. Suitor released the throttle again and soared to around ten feet above the ground, piloting the belt in a loop, and landing safely 16 seconds later.

This was a very encouraging start and, for the moment, Brad Barker and Joe Wright could exchange smiles. But it was only a start. The initial tests, although cautious, had been successful. Now it was time to throw caution to the wind.

The first test flight that would really assess the capabilities of the RB-2000 took place on the following day, 22 January 1995, at Hooks Airport in Houston. The weather wasn't ideal. Grey clouds were gathering overhead, and a breeze whipped around the runway. The area was deserted but for Brad Barker, Joe Wright, Bill Suitor, and two or three friends who had come along to watch and videotape the event. Larry Stanley, not unexpectedly, had not been invited.

The air was full of excitement rather than tension. This was the culmination of a five-year dream. A lot of hard work and an awful lot of money had gone into this project. It is unlikely Barker gave much of a thought to his shattered friendship with Stanley as he helped Suitor, again wearing his white helmet and flight suit, strap on the completed RB-2000.

Suitor took up his position on the tarmac runway, and the small band of onlookers stood back. Then, at 12.32pm, Suitor released the throttle and lifted into the air on a noisy cloud of steam. He soared in an arc over the sea-lane between the runways, creating a huge spray of water beneath him. He reached a speed of 65 miles per hour and an altitude of 30 feet. Then he landed, on tiptoes, having completed a perfect flight of the 'Pretty Bird'. Now he had proved this bird could fly.

'We all jumped up and down and screamed,' said Barker. 'We were pretty pleased.'

Suitor made two further test flights of the RB-2000 that afternoon, but they were somewhat less successful.

'On the last flight he came in a little too hard,' said Barker. 'He rolled over and did some damage to the belt.'

'It damn near killed me,' said Suitor. 'That's when I realised it was a disaster waiting to happen.'

Suitor was lucky to escape uninjured. The rocketbelt was less fortunate, although the damage was repairable. For Suitor, the last test flight highlighted the design problems and reliability flaws he had spotted in the RB-2000. Even Suitor, with years of experience as a rocketbelt pilot, found the RB-2000 difficult to master.

'The RB-2000 had a mind of its own,' said Suitor. 'It weighed over 140 pounds fuelled, and was not easy to carry. The corset was very uncomfortable, and the controls were difficult to move with finesse. It was more like flying a truck with a steering problem.'

Despite these teething troubles, the Rocketbelt 2000 worked. But building the rocketbelt was only the first step. The next step was to make money. It was time to take the RB-2000 into the public domain and set up some paid demonstrations. So Barker and Wright set about repairing the belt, and, within a few months, it was ready to fly for payment.

The RB-2000's first public flight took place at the Houston Ship Canal in June 1995 at a media party organised by Houston Mayor Bob Lanier. The party was thrown in honour of the Houston Rockets basketball team – it had just won the National Basketball Association championship by beating the Orlando Magic. The flight booking was set up by Wright with the assistance of the brother of his former business partner, Bryan Galton. Mayor Lanier's office paid Barker and Wright's newly-named American Flying Belt company $10,000 for the demonstration. Barker and Wright paid Bill Suitor $2,500 to fly the rocketbelt.

Suitor was to take off from a barge on the canal, and Barker and Wright helped him prepare, wearing promotional T-shirts saying 'American Flying Belt. Believe it!' It was early evening, and the setting sun provided an orange glow to the rippling surface of the canal. Police launches darted up and down the waterway, and fireboats sprayed coloured water into the air. At the canal side, a slight breeze whipped at rows of flags and balloons, and hundreds of spectators leaned over railings to get a better view.

Barker strapped Suitor into the belt and asked, 'Too tight?'

'Well, my liver's over here somewhere,' said Suitor, pointing to his chest, 'but that's okay.'

Barker checked his watch, pulled on his protective headphones, and stepped out of the way. Suitor gave the rocketbelt a quick test blast. Then, after a moment of silence, he fired the belt up. To the delight of the pointing crowds, Suitor took off from the barge on a cloud of steam and soared up over the canal, passing a banner reading 'Houston Lighting & Power Salutes The Rockets'. He flew over yachts, with the Loop 610 Bridge behind him, before banking over a group of journalists and making a safe landing in a roped-off area.

It was a perfect flight – for all but one inquisitive TV cameraman. That one member of the large press gathering ducked under safety ropes and strayed into the landing area. He captured some great footage of the rocketbelt coming in to land over his head, but he hadn't bargained for the painfully loud noise from the device's exhausts. Incapacitated by the din, he dropped his camera onto the concrete.

'Then the asshole tried to sue me for hearing loss and damage to the camera,' said Suitor. 'Luckily, the Mayor of Houston called the TV station on my behalf and told them to get the jerk to drop the suit.'

Thankfully, everyone else present seemed to thoroughly enjoy the flight.

'It was nice to see the reaction on everybody's face,' said Barker. 'It was a fun time.'

Barker and Wright helped Suitor remove the belt, and packed it away into Barker's trailer. Then Barker drove away. The first commercial flight of the RB-2000 had been a success. This is what the belt had been designed for – to be flown at paid-for demonstrations. But the public would never see the RB-2000 again. After the Houston Rockets flight, the Rocketbelt 2000 disappeared.

By the time the RB-2000 had completed its first and only public demonstration, Bill Suitor was sick of the whole affair. He had originally planned to fly the belt regularly, but soon changed his mind. He had been living and working with Barker and Wright during the testing of the belt, and the tensions and conflicts between the pair and Larry Stanley had become apparent, and were simply too much to bear. He returned home to Youngstown, and spent much of his time restoring classic cars. The rocketman was now officially retired.

The charges resulting from the hammer attack at Car Audio Plus hung over Larry Stanley and Brad Barker for months. Stanley, now acting as Chief Executive Officer for a technology company called the Microjet Corporation, didn't turn up in court. Instead, knowing Barker would be otherwise engaged, he drove to Car Audio Plus with some friends – including two Harris County detectives – and kicked down the office door in an attempt to retrieve some of the rocketbelt equipment. Then Joe Wright arrived at the scene brandishing the lien paperwork, and the detectives instructed Stanley to leave the equipment. Stanley left empty-handed.

The assault charges against Stanley were eventually dropped, but Barker wasn't so fortunate. He was convicted of Class A Misdemeanour Assault in Harris County Criminal Court Number 11 in December 1995. He was given a suspended sentence of one year in jail plus 80 hours of community service. This was reduced on appeal to a suspended sentence of six months.

In the months that followed, Stanley continued to search for the RB-2000, which had not resurfaced since the Houston public flight. Stanley had seen footage of the flight on a television news broadcast, and that had only increased his desire to reclaim the belt. He had watched the smiling Barker and Wright setting up the RB-2000 and then, for the first time, he had watched the belt fly through the air so impressively. The device he had helped to create worked spectacularly well, and Stanley was determined that Barker would not benefit from it.

Stanley was convinced that the RB-2000 was still in Barker's possession, but its actual whereabouts was unknown. Indeed, Barker himself had also disappeared. He was rumoured to be working on another flying machine project, but no one seemed to know where this work was taking place. One thing that was certain was that Barker had not exploited the interest in the rocketbelt generated by the Houston flight. He had resisted booking further lucrative flights, such was his desire to keep the belt out of sight, and out of Stanley's reach.

Stanley continued to pursue Barker through the courts. His civil lawsuit regarding the ownership of the rocketbelt suffered a succession of major hold-ups that continued for almost two years. During long months of legal wrangles, Stanley accused Barker, Joe Wright, and Jonette Anderson of delaying their pre-trial depositions to avoid having to testify under oath. Barker missed his deposition date several times, and was eventually threatened with jail if he didn't show up. According to Stanley, when Barker did eventually show up, his deposition was contradictory and unreliable. Stanley claimed Barker gave the court the address of a Houston motel as his permanent residence, and then checked out on the following day. Joe Wright admitted in his deposition that proper notice had not been given for Duratron's purchase of ARB's assets. Jonette Anderson claimed attorney-client confidence and refused to answer any questions. Trial was eventually set for 27 July 1998.

While the pre-trial activity continued, Joe Wright's car stereo shop went bust. Business at Car Audio Plus had dropped off. New cars were coming off the production lines fitted with state-of-the-art stereo systems, meaning drivers were less likely to purchase upgrades. Similarly, factory-fitted security features meant car alarms were less popular, and car phones had been superseded by cell phones. Wright had continually strengthened Car Audio Plus by ploughing his profits back into the business, but his involvement with Barker and Stanley had diverted the profits away from the business and into the rocketbelt. Wright estimated that he had lost around $100,000 on the RB-2000 project. He filed for bankruptcy and set about selling the shop's inventory to cover his bills. Wright's dream of running his own business had turned into a nightmare. There was now no way back for the shop or Joe Wright.

Around the same time, Barker and Wright fell out. They had been close friends for over ten years, and the reason for their falling out was unclear. It might have been down to Wright's admittance in court that the Duratron's purchase of ARB's assets had not been fair, or it could have been as the result of a personal disagreement. Barker and Wright had been seen arguing at the workshop, with employees blaming Barker's volatile temper. And Barker had a track record of falling out with friends. With the rocketbelt completed and in his possession, perhaps Barker felt he had no further need for Wright and his workshop. And Wright could not have been happy that the rocketbelt had been removed from the shop despite the effort and cash he had invested. So another friendship was wrecked – but the broken friendship was the least of Wright's worries.

Car Audio Plus finally closed its doors in February 1998. Understandably, there was a change in Wright's formerly sunny disposition. His money was gone, and a lot of his former friends deserted him. Wright spent much of his time sitting in front of his computer, trying to come up with new business ideas. He was very organised, and kept files on everyone and everything on his hard drive. But those files showed Wright was behind on his house payments. And the house was all he had left. He was in an increasingly desperate situation, and sought solace in booze and crystal meth – a powerful form of amphetamines. Wright had fallen hard, and was fast becoming a broken man. The last thing he needed was Larry Stanley's impending lawsuit.

Around the time the audio shop closed, Stanley and Wright began to communicate via a mutual friend. Stanley wanted Wright to help him find the rocketbelt, and Wright wanted Stanley to remove him from the lawsuit. As part of a court-ordered mediation process, Stanley and Wright arranged to meet, along with their attorneys.

'We knew Joe Wright held the key to the rocketbelt,' said Stanley's civil attorney Michael Von Blon. 'Joe had indicated he was willing to help. I don't think he knew exactly where the rocketbelt was, but he indicated he could get us information that would lead us to it.'

According to Stanley, Wright revealed that the rocketbelt was still in the possession of Brad Barker, and that it was located somewhere in North Harris County. Stanley and Von Blon agreed to release Wright from the lawsuit if and when the RB-2000 was recovered, and give him a piece of the rocketbelt business.

'I detected a strong fear in Joe's eyes when he discussed this,' said Stanley, 'and he expressed a great concern that if Barker found that he was negotiating with me he would kill or severely injure him.'

In a further meeting Wright asked Stanley for $10,000 immediately upon recovery of the rocketbelt. Wright planned to leave town. It seemed a sensible thing to do. The Houston dream was over. He had already telephoned his family in Michigan to tell them he would be visiting in August.

'He wanted money up front to get on the road,' said Von Blon.

Stanley claimed Wright's desire to leave town was fuelled by fear. According to Stanley, Wright 'believed Barker would kill him for telling the truth.'

The deal between Stanley and Wright was to be finalised in a meeting at the Houston Greenway Plaza office of Wright's attorney, Ronald Bass, on Wednesday 15 July 1998. Stanley, Von Blon, and

Bass all arrived on time for the 6pm meeting, but there was no sign of Wright. At 6.15pm, Bass telephoned Wright at home. Wright said he was ill and unable to attend, so it was agreed the meeting would proceed with Wright on a speakerphone. There were doubts about Wright's reasons for staying away, and neither Stanley nor Von Blon believed the excuse of illness.

'Joe wanted to help,' said Von Blon, 'but he was scared to death of Barker. Don't forget, Barker had attacked my client with a hammer.'

'I believe Wright was just terrified of being seen by Barker with me,' said Stanley.

According to Stanley, during the course of the meeting, Wright said Barker was a 'psycho' who belonged in a 'little orange suit'. Stanley agreed to pay off the $2,000 mortgage repayment debt on Wright's house, and additionally pay Wright a total of $10,000 upon recovery of the belt. The agreement was drawn up and deemed acceptable by all parties. The meeting ended amicably, and all involved seemed pleased with the outcome.

'I congratulated Joe,' said Stanley, 'and everyone felt like we had reached our goal. Now all that was needed was for Joe to find the rocketbelt for us.'

Stanley and Wright swapped email addresses, and Wright emailed Stanley at 11pm to establish contact.

'I sent him a rather lengthy reply,' said Stanley,

'congratulating him and encouraging him that the future was promising, and suggesting he might be able to help in sales and marketing. I never received a reply.'

Indeed, there would be no further communication at all between Larry Stanley and Joe Wright. Despite Stanley's hopes, and a turnover order signed by the Judge demanding the device be present in court during the trial, Wright never located the rocketbelt. Wright spoke to a friend on the telephone at 11am on the following morning, Thursday 16 July. He was never heard from again.

Three days later, at 3.30pm on Sunday 19 July 1998, officers from the Sheriff's Department were called to Wright's Northwest Harris County home. A friend of Wright, Maurice Heimlich, had found the front door of the house standing ajar, and entered to find a lifeless body lying in a pool of blood on the carpet of the master bedroom. The body had been beaten beyond recognition. The head and torso were completely destroyed. It was impossible to identify the deceased at the scene. It was difficult even to determine whether the body was that of a man or a woman.

Two days later, the Harris County Sheriff's Department were able to make an identification via dental records. Joe Wright was dead. He was 37 years old.

TEN
THE UNSOLVED MURDER

Joe Wright's Colton Hollow home was a modern one-story building with a brick and oak exterior. It boasted an ample garage and a large yard laid with wooden decking. It had clearly been acquired at the peak of Wright's business success. Like most upscale modern homes, it was fitted with several security features. An eight-foot brick wall surrounded the building, security lights were fitted to the exterior walls, and a surveillance camera allowed occupants to vet visitors. None of this had deterred Wright's killer. Indeed, there was no sign at all of forced entry. It seemed Wright had willingly opened the door to whoever had bludgeoned him to death.

Because of the appalling state of Wright's body, it was impossible for the coroner to determine a precise time of death. All that could be said was that Wright had been killed sometime between Thursday evening and Friday evening, the 16 and 17 July 1998, 24 to 48 hours after the speakerphone meeting with Larry Stanley and the attorneys. Although there was very little physical evidence, the brutal method of Wright's murder was clear.

Wright had been struck twice at the door with a blunt object, perhaps a baseball bat. He stumbled backwards and attempted to fend off more blows. He headed for the master bedroom where he kept a gun, but the blows continued. Wright fell onto a rug in the entrance of the bedroom, and the assailant hit him with the blunt object again and again. He was killed by a succession of fierce blows to the head. Investigators estimated that at least 14 such blows had been delivered. Even after Wright was dead, the beating continued. The killer did not stop until Wright's entire upper body had been crushed into a bloody and unrecognisable mess.

Even then the killer was in no hurry to leave, and began to clean up the murder scene, removing all traces of his presence. Even the rug on which Wright had fallen was removed and placed in the bathtub, although it was so completely soaked in blood that the killer made no real attempt to clean it other than turning on the bathwater, which was still running when Wright's body was found. And then the killer left, taking the murder weapon with him. A neighbour reported she might have seen someone climbing over Wright's perimeter wall on Friday evening at around 11.30pm. She could give no further details. Another neighbour reported seeing a black sedan outside Wright's house. The car was never traced.

The running bathwater partially flooded the crime scene and ruined some evidence. Sheriff's Department investigators did take hair and DNA samples, but could not be confident any of them were from the killer. They checked Wright's documents and financial records but did not properly check Wright's computer. Those close to Wright referred to him as a 'file junkie'. He had computer files on everything, they said. His entire life was on his hard drive, carefully filed and organised. Yet the computer was left untouched at the murder scene and eventually turned over to Wright's family. A family member did attempt to retrieve files from the computer but, in the absence of expert assistance, the attempt failed. Other pieces of potential evidence, such as Wright's blood-soaked clothes, the rug from the bathtub and a stash of amphetamines, were also neglected by investigators. Wright's gun, which friends knew he kept in his bedroom, was never recovered.

Maurice Heimlich, the friend who found Wright's

body, said he was sure the horrific scene he discovered would haunt him from the rest of his life. 'I drove over there because I hadn't heard from him in days,' he said. 'As soon as I saw the door was cracked open I knew that something was terribly wrong. I didn't really want to go inside, but I had to see what had happened. There were flies at all of the windows, and I could hear them buzzing. There was a horrible stench. It was one of Houston's hottest summers – I think it had been 110 degrees for three months straight that year. As I walked in I saw a spot or two of blood on the floor on the way to the bedroom. Then I saw the body lying on the floor. It was pretty bad. I didn't examine the body or anything, but I knew it had to be Joe. It was by far the worst day of my life.'

Although it was outside of their jurisdiction, the FBI offered to help with the case. They were intrigued by the rocketbelt, and saw the murder as a high profile case. It would soon become clear that the FBI also had another interest in the case. But the Harris County Sheriff's Department turned the offer down. Wright's family and friends were unhappy with that decision, and with the Sheriff's Department's handling of the case.

'The police screwed up the crime scene,' said Nancy Wright, Joe Wright's younger sister, who pressured police to find his killer. 'They didn't collect any good evidence.'

But the Sheriff's Department did have a suspect. The murder had taken place just a week before the scheduled civil trial, and detectives could not ignore the rocketbelt connection. Within a few weeks of the murder, Brad Barker was arrested in a Houston pawnshop.

Barker said he first heard of the murder when he received a phone call from his ex-wife while in Arkansas on the night of Sunday 19 July.

'She called me at my office in Fort Smith,' he said. 'I remember specifically the call. It was probably eight o'clock at night. She said, "Joe Wright was murdered." I said, "You're shitting me."'

Barker's ex-wife told him she had heard the story from Larry Stanley's wife. That made Barker suspicious. There had been nothing in the newspapers, and he knew Stanley was desperately trying to track him down. Perhaps it was a ruse designed to draw Barker into the open.

'As soon as I got off the phone, I called a couple friends of mine in Houston and told them what I had heard,' said Barker. 'I asked them to look in the paper. After a couple days they still hadn't heard anything about it, so I just completely let it go.'

In fact, brief notice of Wright's death did appear in the *Houston Chronicle* on 21 July 1998 under the headline, *Slain man identified.* In a 50-word piece, the newspaper reported that, 'a man found beaten to death in his northwest Harris County home has been identified as Joseph Wright, 37.'

But the small article was easy to miss, and Barker's friends didn't see it. Barker himself didn't seem too concerned about his former friend. He didn't take any further steps to verify the rumours of Wright's death. It seemed he was happy to assume that the rumours were false.

'Then,' said Barker, 'about a month later I was in Houston and got taken in by the Sheriff's Department.'

Barker underwent questioning and was held for three days. He claimed he had not spoken with Wright for over a year, and had been in Fort Smith at the time of the murder. According to Barker, telephone records and the testimony of friends in Arkansas could verify his whereabouts. Sheriff's Department investigators took hair and DNA samples from Barker, and also confiscated personal items, including his car. Barker claimed his Pontiac Grand Am was taken apart and destroyed.

'They held me for 72 hours on what's called investigative hold,' Barker said. 'They can hold you that long without arresting you. Then they were finally forced to release me, and they let me go.'

Barker vehemently denied any involvement in the murder of Joe Wright, but, despite the fact that he was released without charge, Sheriff's Department officials continued to name him as the only suspect in the case. Barker complained about his treatment at the hands of the Department, which he saw as completely unwarranted.

'The Sheriff's Department has gone out of its way to implicate me in this murder,' he said, 'and I've gone out of my way to prove I am innocent.'

During questioning, Barker gave investigators a list of names of people he said could verify his being in Fort Smith at the time of the murder. He was adamant they could clear his name. Barker also claimed telephone records would prove he made a number of phone calls from Fort Smith at that time. Upon his release, Barker meticulously collated as much information as he could to back up his alibi.

'I took all the records to the Sheriff's Department,' said Barker, 'and they more or less laughed at me and wouldn't even look at them.'

Detectives said they did contact every person on Barker's list, but could not substantiate his story. Despite the fact that there did not seem to be any trace of evidence linking Barker to the murder, the Sheriff's Department refused to eliminate him from the investigation. Brad Barker remained a suspect.

But did Barker have a motive to kill Joe Wright? The pair had fallen out, and were not on speaking terms. Wright was planning to help Larry Stanley locate the rocketbelt. Barker could not have been happy with that, and may have viewed it as a betrayal. Wright had spoken to friends and his attorney about his fear of Barker. He had mentioned

Barker had left threatening answerphone messages at his home. Detectives found these messages and listened to them, but decided Barker seemed angry rather than threatening.

Barker said he had been angry because Wright had given a friend's telephone number to the FBI in relation to a copyright infringement dispute involving the rocketbelt. 'I remember calling Joe Wright,' said Barker. 'I didn't get him, I got his answering machine, and I remember specifically, I said, "Hey Joe, it's your buddy Brad. You need to stop blaming other people for your faults and get on with your life." That was it.'

The timing of the murder suggested it could have been related to the speakerphone meeting on the Wednesday night. Stanley and the attorneys believed Wright had not shown up because he had been frightened that Barker would find out he was colluding with Stanley to locate the rocketbelt. But how could Barker have found out about the meeting? Nancy Wright believed Stanley might have told Barker, or an associate, that Wright was helping him in an effort to annoy him.

'Larry has a big mouth,' Nancy Wright said. 'Larry probably put the word out that Joe had agreed to go against Brad, hoping it would get back to Brad and piss him off.'

Stanley's attorney Michael Von Blon had a different theory. 'I believe Barker followed Stanley to Ron

122

Bass's office,' said Von Blon. 'He thought, "What are Stanley and Von Blon doing here?" and he put two and two together.'

Maurice Heimlich confirmed that Wright had been afraid of Barker. 'For the last month, especially in the last two weeks, he was in constant fear for his life,' he said. 'He was 100 percent sure that Barker was going to try to kill him.'

So, according to the Sheriff's Department, Barker was a suspect. But what of Larry Stanley? He had no clear motive to kill Wright, and claimed to be on good terms with him following the speakerphone meeting. Nancy Wright said she asked the Sheriff's Department to give Stanley a polygraph test, but they refused. He was never questioned by detectives in connection with the murder. Stanley had friends in the Sheriff's Department, and he publicly exerted pressure on them to arrest Wright's killer. Stanley said he was in no doubt over the killer's identity.

'I am absolutely convinced that Bradley Wayne Barker murdered Joe Wright at Wright's home on 16 July 1998 with malice aforethought, by beating him to death in a furious rage with a baseball bat or other blunt instrument,' Stanley said in his affidavit. 'I believe Barker is a serious threat to my person, my family and society, and that he will not rest until he murders me and anyone around me at some future encounter. The sooner Barker is in custody the better off my family and society will be.'

The fifteen-page affidavit was posted on Stanley's Microjet Corporation website, in anticipation of the civil trial, shortly after Wright's murder. It was illustrated with colour photographs showing the rocketbelt and the various people involved with it. He also included the bloody photographs taken in the aftermath of the hammer attack.

Stanley met with Wright's family and friends, and shared his belief with them that Barker was the killer. The family claimed he told them that, if law enforcement failed to punish Barker, he would take care of it himself. He had friends, he allegedly said, who would kidnap Barker, take him out into the desert, torture him, and kill him. Wright's family was shocked, although they didn't believe that Stanley could be capable of such a deed. Nevertheless, Stanley fervently pushed for Barker's prosecution.

'Larry was insistent that Brad had killed Joe,' said Nancy Wright, 'and he wanted to see he was punished.'

Stanley had already made clear in his affidavit that he was afraid of Barker, and a former employee of Wright had warned Stanley that he might be in grave danger. So when Stanley reported that Barker had driven past his home, the Harris County District Court issued a restraining order forbidding Barker from coming within 300 feet of Stanley. Barker claimed he had not driven past Stanley's house, and had been in a Dallas church with his mother at the time. He said his signature in the church register would confirm that.

Stanley became more concerned for his own safety, and that of his family, when his son spotted a man in the backyard of their Sugar Land home. He obtained a concealed weapons permit and bought a .40 calibre Desert Eagle pistol that he carried at his side at all times for protection. He showed the gun to Nancy Wright when she visited to discuss her brother's murder. She was sitting in a lawn chair in his garden as he waved his new possession about. Then the gun accidentally discharged and almost killed her.

'I felt the heat of the bullet whiz by my head, and had trouble hearing out of my left ear for a couple days,' she said. 'The bullet took a fist-sized piece of sod out of the yard. Larry called the local police to tell them what had happened in case any of the neighbours complained. He had that gun on him everywhere he went. He was convinced that Brad was out to kill him.'

It seemed Stanley genuinely feared his former friend Brad Barker. Certainly, Joe Wright had feared Barker. But Wright had also feared someone else.

In the initial days of the investigation there were two other suspects in addition to Barker, both of whom had known Wright well. Just days before his death, Wright told a friend he feared for his safety after a disagreement with local bookmaker Mike Bowman. Wright had drawn up a file on Bowman, and told the friend that, should anything happen to him, the file, which was hidden in Wright's attic, should be handed to the police.

Wright didn't gamble, but he became friendly with Bowman in the early 1980s. Bookmaking is illegal in the state of Texas, and Bowman was a convicted criminal, but he also ran a legitimate business – a strip club – on the Houston mall next to Car Audio Plus. Wright bartered and exchanged services with Bowman, as he did with many friends and associates. Bowman helped Wright with his financial accounts, and Wright bought Bowman a car. When Car Audio Plus began to struggle financially, Bowman loaned Wright $30,000. In return, Wright signed over his $50,000 life insurance policy to Bowman. After a 15-year friendship, just two weeks before Wright was killed, the pair fell out. According to friends, Wright, in desperate financial trouble, asked Bowman for more money and was refused.

After Wright was killed, friends pointed the finger of suspicion at Bowman. One claimed that Wright had told them he was afraid Bowman would have him killed, but detectives found nothing to connect Bowman to the murder. The unusual life insurance policy did raise a red flag at the insurance company, but the Sheriff's Department had Bowman take a lie detector test, which he passed. The policy was paid out in full.

As for Wright's file on Bowman, the contents were never revealed. Sheriff's Department detectives checked the file out, but claimed to find no evidence relating to the murder. According to Brad Barker, who looked into the whereabouts of the file, when the District Attorney's office began to investigate the case they found that the file had disappeared from the Sheriff's Department evidence room. Suspicions were raised, and rumours began to circulate that the file implicated members of the Sheriff's Department in the operation of an illegal gambling ring run by Bowman. Friends said that Wright was close to Bowman, and to a Lieutenant in the Sheriff's Department, so, if such an activity took place, it was entirely possible Wright would have known about it. But, without Wright's file, the allegations were nothing but hearsay.

Then Wright's friends pointed to one of Bowman's employees. Wayne Johnson, a small-time drug dealer, sold Wright his crystal meth. Johnson had police records for assault, and for possession of illegal firearms and drugs. He'd been a visitor to Car Audio Plus during the construction of the rocketbelt, and had been hired by Wright to work security at the Houston Ship Canal flight. The pair had been friends, but had recently fallen out over a money loan, perhaps because a desperate Wright didn't have the cash to pay for the drugs he craved. It was known that Johnson had threatened Wright over the loan.

Both Bowman and Johnson denied killing Wright, but others maintained that the pair could both have been involved. Friends of Wright suggested that Johnson could have carried out the murder on behalf of Bowman. Although no evidence was found to link him with the murder, a close relation of Johnson was convinced he was the killer, and revealed that suspicion to Wright's family.

The FBI didn't agree. At this point they revealed their interest in the murder. They were keeping a close eye on Bowman and Johnson as part of an investigation into illegal gambling. And they eliminated both men as suspects in the murder of Joe Wright. Harris County Sheriff's Department, however, held a differing opinion. They cleared Bowman, but retained Johnson as a suspect. However, perhaps surprisingly, Johnson was not officially questioned.

Another intriguing angle of the investigation centred around a woman named Diane. A friend said that Wright had mentioned he had a dinner appointment with Diane on the Thursday evening around which he was killed. None of Wright's friends or family members could say for sure who Diane was, although some suggested she could have been a crystal meth dealer. When Wright's body was found he was dressed to go out in a smart shirt and blazer. It seemed likely he had been ready to head out to meet Diane, or was expecting her to arrive at his house. Detectives were unable to locate Diane, or confirm who she was. Could she have been the killer, or could the killer have used her to persuade Wright to open his door?

Another individual who was named by associates of Wright in relation to the murder but was never investigated was Bryan Galton, Wright's former business partner. Galton left Houston in 1986, but returned to the city some months before Wright was killed. Friends claimed Galton renewed his friendship with Wright, but things soon turned sour. They accused Galton of beating up Wright on more than one occasion. Friends also suggested that the relationship between Galton and Wright extended beyond that of business partners.

Wright was a secretive person, and he kept even the most important aspects of his personal life hidden from all but a few confidants. He was gay, and he was very uncomfortable with that part of his life. Wright didn't lead an openly gay lifestyle, and kept up a façade of being straight. He found his homosexuality difficult to accept, and even harder to reveal. Only Wright's closest friends knew he was gay. Wright had not even been able to tell his father. Those friends who did know about Wright's sexual orientation felt it had affected the Sheriff's Department's handling of the case.

Nancy Wright said detectives made insulting comments in her presence about Wright's homosexuality. 'One detective insinuated that Joe had a preference for young men – teenagers – and that his house was full of homosexual pornography,' she said. 'Another said that he did not condone the gay lifestyle and felt it was morally wrong. I told him it was not up to him to pass judgment on people's lifestyles, and that he was paid to serve and protect the public and should probably keep his bigoted remarks to himself.'

The fact that Wright was gay might have been crucial to the investigation. The brutal manner of Wright's murder, with multiple blunt trauma wounds inflicted after death causing body mutilation, is referred to by murder investigators as 'overkill'. Far more blows were used than were necessary to kill Wright. This type of violence suggests rage from the killer against the victim, and therefore implies a relationship between the two parties. Overkill is almost never associated with random killings. Instead, it is associated with personal killings and so-called crimes of passion.

So the manner of the murder suggested Wright had been killed by a close associate over a personal matter. But the close associate was not necessarily a lover, and the personal matter was not necessarily related to a personal relationship. Those closely involved alongside Wright with the RB-2000 had become dangerously obsessed with it. Could their obsession with the rocketbelt have led to murder?

ELEVEN
THE $10 MILLION LAWSUIT

In early 1999, six months after he was questioned over the murder of Joe Wright, Brad Barker found himself back in the custody of the Harris County Sheriff's Department. He was driving south of Houston when he spotted flashing lights in his rear view mirror.

'I'm not breaking any laws, I'm not speeding, I'm not doing anything,' he said. 'Next thing I know I've got cops all over me.'

Barker pulled over to the side of the road, with the police car tucking in behind him. He waited at his steering wheel and wondered what he had done wrong this time. Armed Deputies approached Barker's car and asked him to identify himself.

Barker offered his name. Then one Deputy said, 'Get out of the car, Mr Barker. There's a warrant for your arrest out of Houston for the theft of a firearm.'

'You're full of shit,' said Barker. 'I don't know what you're talking about.'

What the Deputy was talking about was a warrant relating to an alleged theft reported by one Larry Stanley. Barker was accused of stealing Stanley's Winchester rifle from the Car Audio Plus office almost five years previously in November 1994. Barker was arrested and held pending a trial. He spent 60 days in jail.

On 12 April 1999, Barker and Stanley finally came face to face, in the 338th Judicial District Court of Harris County. At the Courthouse, in a downtown Houston high-rise, Barker admitted taking the rifle from Car Audio Plus, and subsequently pawning it for $100. But Barker claimed the rifle was his, and had been given to him by Stanley in June 1992 as collateral to cover a $75 cheque that had bounced. Stanley said that would have been impossible, as he had not bought the rifle until December 1992.

District Court Judge Elsa Alcala was unimpressed by either party. The Judge said both Barker and Stanley had reason to lie because of their feud over the rocketbelt. She said Stanley had not given straight answers in court, and could not prove ownership of the rifle. She also said Barker could not prove he had received the rifle as collateral for the bad cheque. Judge Alcala decided there was reasonable doubt, and Barker was acquitted.

But this was hardly a victory for Brad Barker. After being held for three days without charge after the murder of Joe Wright, Barker had now been held for a further 60 days on a charge of which he was acquitted. The loss of freedom was particularly distressing to Barker as it meant he was separated from his young son. But Barker's suffering was set to increase.

Larry Stanley's civil suit finally came to trial on 26 July 1999 in the 269th Judicial District Court of Harris County, on the third floor of the Houston Courthouse. Since the suit had been filed in 1995, the trial had been delayed by parties failing to give depositions, refusing to be interrogated, and failing to present evidence. Then Joe Wright's murder in July 1998 had necessitated the abandonment of that month's court date. In total, Judge John T Wooldridge was forced to reset the trial date six times. By the time the case came to court, with Wright now dead, the title of the suit had been amended. It was now: Thomas Laurence Stanley III, on behalf of the American Rocketbelt Corporation, versus Bradley Wayne Barker, the estate of Joseph Wright, Duratron Inc, and Jonette Anderson.

Larry Stanley turned up at court with his civil

attorney Michael Von Blon, and Von Blon's associate Kirsten Davenport. Also present was Stanley's brother, Jerry, who was acting as attorney for the American Rocketbelt Corporation. Jonette Anderson, accused of setting up Duratron to defraud ARB, arrived to represent herself. She asked for a summary judgment in her favour, but the court found her delaying and obstructive behaviour in the pretrial process had abused the court process. After her request was refused, Anderson walked out of the court. Joe Wright's estate was not represented. Brad Barker, who had represented himself during the pre-trial, failed to show. The RB-2000, which the Judge had ordered should be present in court, was similarly absent.

Proceeding with none of the defendants present, Michael Von Blon outlined Stanley's case. He contended that the defendants had seized the rocketbelt and associated equipment through fraud 'for their own personal use, benefit, and monetary gain'. He said Brad Barker and Joe Wright had drawn up a false lien alleging ARB owed back payments of rent to Wright. Then Jonette Anderson had set up a fraudulent company, Duratron, to purchase ARB's seized assets for just $10,000. Von Blon stated he was seeking the return of the rocketbelt, plus damages of $2 million to $3 million, equating to an estimated $500,000 per year in lost earnings since the theft. Von Blon also alleged that Barker had assaulted Stanley with a hammer at Car Audio Plus in November 1994. The photographs taken at the time of Stanley's bloodied head were presented as evidence.

Then Stanley took the stand and explained what the rocketbelt was, how it worked, and how much it could earn. The court was informed that the James Bond-style rocketbelt was a lot like that featured in *Thunderball*. Stanley told the court that other rocketbelts were known to earn $25,000 plus expenses per flight. When asked why Barker would want to steal the belt, Stanley replied, 'He worshipped that belt. It was his golden calf. He considered it his personal property. He would polish it all day long. It was an obsession.'

With no defendants or defence attorneys on hand to present a case for the defence, the trial lasted just a day and a half. Stanley and his attorneys had offered up a strong case, without any opposition. They must have been confident of a positive outcome, but they would be stunned by the manner of it.

On 27 July 1999, Judge Wooldridge presented his findings. In his summary, the Judge said the case contained, 'evidence of greed, craft, intrigue, mystery and murder – elements of a cheap novel found on the bestsellers list.'

The court found the defendants had individually committed fraud against Stanley and ARB, and entered into a conspiracy, in an attempt to procure the rocketbelt. The court also agreed that the rocketbelt was capable of generating an income of $25,000 per flight, and $250,000 to $1 million per year, over an expected lifespan of 20 years. Judge Wooldridge ordered that the defendants pay ARB and its shareholders $7.5 million for loss of earnings, and $2.5 million in further damages.

The court also found that Brad Barker had committed an assault on Larry Stanley in November 1994, and then filed a malicious prosecution falsely accusing Stanley of assault. Judge Wooldridge ordered that Barker pay Stanley $111,766.25, representing damages of $10,000 for assault, $1,766.25 for medical expenses, $50,000 for malicious prosecution, and $50,000 for disfigurement resulting from the assault. It was also decreed that the defendants should pay Stanley $131,047.51 in legal costs.

Finally, the court found that Barker had breached the terms of his contract with Stanley. Therefore, it was ordered that the defendants or their associates turn over the rocketbelt and its associated parts to Stanley, and that the defendants refrain from damaging, selling, disposing of, or diminishing the value of the rocketbelt.

In total, Stanley was awarded $10,242,813.76, plus sole ownership of the RB-2000. The massive award was far in excess of his and his attorneys' expectations, but would he ever see either the money or the belt? Jonette Anderson declared herself to have no assets, Joe Wright was dead and his estate was empty, and Brad Barker could not be located. It seemed to be a hollow victory, but Stanley was optimistic the verdict could precipitate the return of the belt.

'I'm relieved this is over with,' Stanley told reporters as he left the court. 'Hopefully this will exert public pressure on them to finally return the belt.'

Barker, meanwhile, said he hadn't been aware of the trial until he read about it in a newspaper. 'The reason I wasn't in court was because they were sending letters to me at Fort Smith when they knew I was in Dallas,' said Barker. 'And Larry Stanley was having one of his guys up there take my mail and not give it to me. So, basically, he walked into the court of law without me there to defend myself, and the Judge awarded him $10.2 million.'

Barker didn't pay too much attention to the size of the judgment against him. 'I wasn't really worried about losing $10.2 million since I didn't really have it,' he said. 'I just kind of laughed at it. They should have just asked for $100 billion.'

He dismissed suggestions of returning the belt to Stanley, and refused to say whether it was in his possession. 'Even if I had it,' he said, 'I would smash it into a million fucking pieces with a road grader.'

Such talk did not dissuade Stanley, who vowed to

continue his search for the belt, offering a $10,000 reward for its recovery. 'If I'm persistent and determined, I will recover that belt,' he told court reporters. 'I'm not anywhere close to giving up on it. He is going to have to bury it.'

Since the disappearance of the RB-2000, Brad Barker's whereabouts had been largely unknown. He had been living variously in Houston, Dallas, and Fort Smith, and working on a secret project, the details of which were now revealed. Barker had been working on a new flying machine with the financial help of Vinson Williams, the owner of the Williams Tool Company, a supplier to the oil industry.

The new device was a flying hovercraft shaped like a giant beer can. Barker called it the Personal Flying Device (PFD). Barker and Williams planned to sell the idea to beer companies and soft drinks manufacturers as a promotional tool. The idea was that the device would be painted in livery that matched the sponsors' drinks cans. Barker envisaged giant Pepsi and Budweiser cans flying through the air at promotional events. Williams handed over an initial loan payment of $57,000 upon seeing the business plan, and promised serious financial backing for the project. Barker built the PFD in Houston with colleagues including another friend from his past, Tom Wade.

Barker had even begun to mend old rifts, and had rekindled his friendship with Kinnie Gibson. Gibson was now running a company called Powerhouse Productions, continuing to fly the Tyler Rocketbelt, and trading under the name Rocketman Inc. Gibson bought Barker a Bible as a gift, and Barker said he read it regularly. But had Barker learned his lesson, and decided never to steal from his friends again? Possibly not, as the idea for the PFD was very similar to that of Barker's RB-2000 collaborator Doug Malewicki, who had invented a 'flying can blimp' some years previously.

While the friendship with Gibson was repaired, another fell apart. The PFD failed to get off the ground – literally. The contraption simply didn't work. Production moved from Houston to the Williams Tool facility in Fort Smith, but fortunes didn't improve. Barker and Wade argued, and costs spiralled. Then the murder of Joe Wright, and the subsequent accusations surrounding Barker, understandably made things very difficult. Finally, a cash-strapped Barker began to pawn some of the PFD parts. It was clear that the PFD, unlike the RB-2000, would never be able to fly.

The failed venture had cost Vinson Williams an estimated $400,000, almost ten times his initial investment. He swiftly pulled the plug on the project, and locked Brad Barker out of the Fort Smith shed in which the device was kept. Barker had no other means of income, and he was insistent that Williams should allow him to have the failed PFD. He claimed that papers drawn up between the pair proved his entitlement to the device, and that Williams was appropriating the PFD and Barker's business plan for his own use.

'I tried to negotiate with the guy to get it back, but he wouldn't even talk to me on the phone,' said Barker. 'So, like an idiot – and I take full responsibility for being an idiot here – I went back to Fort Smith.'

On the night of 1 September 1999, Barker and Wade arrived in darkness at Williams Tool with a set of walkie-talkies. The desperate Barker had formulated a risky plan. With Wade acting as lookout, Barker approached the locked shed that housed the PFD and climbed up to the building's air duct. He squeezed through the narrow gap, and managed to crawl into the dark shed. But, unbeknown to Barker, an aggrieved Wade had informed Williams of the plan, and Williams had informed the authorities. A group of armed Sheriff's Deputies were waiting in the shed. As Barker slid through the air duct, the Deputies flipped on the lights and caught him in the act. Barker was wrestled to the ground with several guns pointing at his head. More Deputies rushed from hiding places in nearby cars and trucks. Barker had been set up. He was arrested and charged with commercial burglary.

THE $10 MILLION LAWSUIT

So Brad Barker found himself behind bars for the third time in less than a year, and his stay in the Sebastian County jail would be his longest period of incarceration yet. He had never been in trouble with the law before his involvement with the Rocketbelt 2000, but now he was becoming very familiar with the inside of a jail cell. On this occasion, the District Attorney did offer to drop the burglary charge if Barker admitted to the lesser misdemeanour charge of criminal trespass, but Barker refused.

'I said, "Piss on you. I'm not pleading guilty to anything. That was my equipment. I got screwed out of it,"' he said.

Barker spent a further 72 days in jail. Bail was initially set at $100,000, but Barker managed to get this reduced to a more affordable $25,000 so he could travel to Houston to retrieve papers he claimed would clear his name. He said the papers, signed by Williams, would prove he had been falsely arrested for breaking into his own shed. Barker raised the bail bond and was released, but, when he arrived at his storage facility in Tomball, Texas, it was empty. It seemed Larry Stanley had beaten him to it.

Within hours of hearing of Barker's arrest, Stanley had hurried to Fort Smith in the hope of locating the rocketbelt. He met up with Barker's former PFD colleague Tom Wade, who led him to the hiding place of the rocketbelt fuel laboratory trailer. But the rocketbelt itself was nowhere to be found. Stanley took possession of the trailer, anyway. Then Wade showed Stanley to Barker's Toyota. Stanley apparently searched every inch of the vehicle, finding Barker's address book and a receipt relating to the lease of a storage facility in Tomball. He immediately headed back to Texas. Could Tomball be the hiding place of the RB-2000?

Stanley had no trouble gaining access to the storage facility. Barker had given Tom Wade the keys during his incarceration and asked him to retrieve the signed agreements he said would clear his name. Instead, Wade gave the keys to Larry Stanley. Stanley opened the lock-up, but he didn't find the rocketbelt. The golden goose was not there. Instead he found Barker's paperwork.

'Basically,' said Barker, 'he broke in and stole all of my documentation.'

Barker reported the burglary to the Harris County Sheriff's Department. He had no doubt who was responsible. Certainly not after Stanley sent a fax to the house owned by Barker's mother. According to Barker, the fax told him not to try looking for the papers, because he wouldn't find them. Stanley had them, and they were available – for a price. If Brad Barker wanted to reclaim the papers that he believed would clear his name, he would first need to hand over the Rocketbelt 2000.

TWELVE
THE KIDNAPPING

In November 1999, Brad Barker began to receive messages from a Hollywood stuntman named Chris Wentzel. Chris 'The Flying Wizard' Wentzel specialised in airborne stunts, and had featured in a bunch of not very successful movies, the biggest of which was 1992's *Poison Ivy*, starring Drew Barrymore. Wentzel said he wanted to offer Barker a three-day movie job in the desert near Los Angeles, filming the controlled crash of an airliner, and paying around $1,500. Wentzel's messages were passed to Barker via a mutual acquaintance named Stan Casad. Wentzel had met Casad, who worked within the shady fringes of the medical industry, while hunting for treasure some years previously.

Barker, who had only been out of prison for a few days, wasn't particularly keen to take the job, but he did need the money. He had also, of course, held an interest in movie work ever since watching his friend Kinnie Gibson become a Hollywood stuntman. Barker was still on bail following his release, so he called his bail bondsmen, J&J Bonding in Fort Smith, and left a message to inform them that he was heading to Los Angeles for work and would be back in a few days. On 26 November, Barker left Arkansas and travelled to California.

Almost as soon as Barker had crossed the state line, Larry Stanley began to call J&J Bonding. He told the company that Barker was jumping bond and planning to leave the country for Venezuela. Stanley also persuaded a Fort Smith Sheriff's Deputy and Detectives from Harris County to warn the bond company that Barker was absconding.

Unaware of the furore back in Arkansas, Barker landed in Los Angeles and was met at the airport by Wentzel, a short, stocky man with a thick neck and short dark hair. Wentzel took Barker out for a meal, and the pair then took a ride on Wentzel's boat, before heading back to his North Hollywood home. Outside the small white house stood two men Barker didn't recognise.

'These are the guys you'll be working with in the desert,' said Wentzel.

Barker introduced himself, and all four men went

inside. They sat at a kitchen table and chatted for around 15 minutes. The atmosphere was extremely friendly, and Barker began to look forward to working with the men.

Then Wentzel pulled out a gun and pointed it straight at Barker's head.

Thinking Wentzel was kidding around, Barker smiled. 'I thought it was a joke,' he said. 'I wasn't being cocky or anything, but when I smiled it scared the shit out of Chris Wentzel.'

Scared or not, Wentzel erupted in fury. It immediately became clear that he wasn't joking. 'Get that son of a bitch on the ground!' he yelled. One of the other men grabbed Barker in a chokehold and threw him to the kitchen floor. 'The last time I put a gun to somebody's head and they smiled at me,' said Wentzel, 'I knew they were crazy or they just didn't give a shit.'

Barker stopped smiling. 'At that point I realised it was serious,' he said. Within seconds, a convivial meeting between co-workers had transformed into a terrifying violent encounter. But why? Barker was sure he had not said or done anything to upset these men. So why was he being held down on the kitchen floor with a firearm pressed to his head?

Wentzel stood over Barker, brandishing the gun. 'Where's the rocketbelt?' he demanded.

Barker was shocked. What did the rocketbelt have to do with any of this? Chris Wentzel had no involvement whatsoever with the RB-2000. As far as he was aware, Wentzel did not even know the rocketbelt existed. Then he had a sudden realisation.

'So Larry Stanley's behind this,' Barker said.

'Don't know him,' said Wentzel. 'Never heard of him.'

Then Wentzel and the others set about tying Barker up. The men wrenched Barker's arms behind his back and fixed him in handcuffs. Then they sat on his legs and tied them together with rope. They pulled him into a seating position, and Wentzel pressed the gun to the back of Barker's head.

'Did you kill Joe Wright?' asked Wentzel.

'No,' said Barker.

'Where's the rocketbelt?' said Wentzel.

'I don't have it,' Barker answered.

The interrogation continued for several hours. Wentzel was determined Barker would confess to killing Joe Wright and give up the whereabouts of the rocketbelt. Eventually, Barker admitted that he did know where the rocketbelt was located. He said a friend was holding it as collateral in a loan he had taken, but he refused to say any more. Then Barker had a hood duct-taped over his head, and was carried into another room. There he was pushed into a small white wooden box, around three feet high and four feet long, with the words 'SCUBA TANKS' stencilled on the side.

'Watch your fucking head, bad boy,' said Wentzel as he pushed a lid over the box, forcing Barker down into the cramped box. Then Barker heard the sound of a power drill. Wentzel was screwing the lid shut.

Barker now found himself handcuffed, tied, hooded, and trapped in the sealed box. And he would remain there for several days. The Rocketbelt 2000 had been meant to allow Barker to soar through the air like a bird. Now his involvement with the device had left him caged in a wooden box. The RB-2000 had caused broken friendships, violent confrontations, accusations of murder, a multi-million dollar lawsuit, and now a very odd kidnapping. Surely Barker was now beginning to regret ever deciding to build a rocketbelt.

Each day, Wentzel quizzed Barker about the whereabouts of the RB-2000. Wentzel told Barker they would kill him, and harm his son, unless he complied. 'I know exactly where your son lives, and I'll go get him and tie his little ass up in the box with you,' he said. Wentzel also asked Barker, menacingly, if he was scared of rats or snakes. The interrogation continued relentlessly, but Barker said nothing. He figured the men were going to kill him anyway, and the only thing keeping him alive was the fact that they needed information from him. He wasn't just trying to protect the whereabouts of the rocketbelt. He was trying to stay alive.

Barker spent his time in the box trying to figure

out what was going on. His captors wanted the rocketbelt, and they also wanted him to admit to killing Joe Wright. It was obvious that the job offer had been a ruse to lure him out to Hollywood. But what was the connection between Wentzel, the rocketbelt, and Wright? It seemed that Wentzel was receiving instructions and questions via email. Who else was involved? Wentzel had named Barker's son, and correctly stated where he lived, but Barker had never mentioned his son to Wentzel. Where did Wentzel get that information from? Barker was convinced that Larry Stanley was somehow mixed up in this, although Wentzel had denied that was the case. So who were the emails from? And why did Wentzel constantly refer to Barker as 'bad boy'?

During one interrogation, a desperate Barker suggested Wentzel call in Kinnie Gibson to mediate. Wentzel related this request to the mystery emailer, who responded with the message: 'Bad boy suggests we get Gibson involved? Since Gibson is the stunt double for a Hollywood tough guy, maybe little bad boy hopes Gibson will come out here and rescue him from the big bad men holding him.' The emailer also threatened to force his way with a gun into the home of the person who was holding the rocketbelt in an attempt to capture the device.

At one point in his ordeal, during which he lost all track of time, Barker was pulled from the box with the hood still taped over his head and moved to another room.

'My friend Jim is helping me today, bad boy,' Wentzel said of a second captor. All Barker could see through a gap in the hood was that Jim was wearing white tennis shoes.

Barker was carried into Wentzel's garage, but if he thought his nightmare was coming to an end he was to be sorely disappointed. His captors dragged the empty scuba tank box into the garage, picked Barker up, and stuffed him back inside. They pushed him down and reaffixed the lid with the power drill. But after they had screwed it down, Barker could still hear the sound of the drill. They were drilling holes in the box.

'Is that enough holes?' asked Jim, in a vaguely familiar voice.

'No,' said Wentzel. 'The more holes, the faster it'll sink.'

Barker began to weep. It seemed his captors were planning to dump him in the ocean.

'Please,' he begged. 'Let me out!'

'Shut the fuck up,' said Wentzel. 'We're going for a little boat ride.'

'Please, Chris!' cried Barker, 'I don't want to drown. Would you give me a minute to say a prayer, and then put a bullet in my head?'

'Sure,' said Wentzel. 'No problem.'

Barker said his prayer and waited, but the bullet never came, and neither was the box dumped in the ocean. Barker stayed in the box, and the lid was lifted only twice – to feed him a single slice of pizza, and a small cup of soup. That wasn't enough to prevent Barker from being starving and thirsty. He was cramped and aching, and the handcuffs were cutting into his wrists.

Then, after six days, Wentzel opened the box and pulled Barker out.

'I've got a notary coming by,' said Wentzel, 'and you are going to sign some papers. If you look at her, if you try to talk to her, I'll blow your head off.'

Barker was dragged into the kitchen and one of his hands was handcuffed to a chair. Then the hood was torn from his head. Once Barker's eyes adjusted to the light, he saw that sitting opposite him was the notary – later identified as Wentzel's friend, Elyse Hoyt. On the table was a sworn statement waiving Barker's claim to the rocketbelt, plus papers relating to his bail extradition back to Arkansas. Wentzel pointed his pistol at Barker's head.

'You are going to sign this thing or else I'm going to shoot you in the head,' he said.

Wentzel pointed to the places on the waiver that required signatures, and Barker duly signed with his uncuffed hand. Hoyt notarised the papers and left. Then Wentzel opened the door.

'I think it's time you met Jim,' he said.

And in walked Larry Stanley. Barker was dumbfounded. Stanley was wearing the white tennis shoes Barker had spotted earlier, and was brandishing his Desert Eagle pistol. He sat opposite Barker and placed his gun on the table.

'Where's the rocketbelt?' asked Stanley.

Barker refused to answer.

'If you don't cooperate, things could get worse for you,' said Stanley.

Barker said nothing.

'You know I took your paperwork from Tomball?' said Stanley.

Barker remained silent.

Wentzel gathered up the signed documents from the table and handed them to Stanley. Stanley took them, along with his pistol, and left. Then Wentzel released Barker from the chair, reaffixed his handcuffs, and put him back in the box.

It was obvious now that Larry Stanley had been behind the kidnapping from the very beginning. He had recruited Wentzel and the others, set up the Hollywood trap, and kept in contact with the kidnappers via email. Stanley had gone to cruel and unusual lengths in an effort to retrieve the rocketbelt. And that made Barker even more determined not to give it up to him.

The next day, his seventh in captivity, Barker was pulling on his handcuffs when one of the cuffs popped loose by a notch. He continued to pull, and managed to pop more notches. Eventually, the cuff fell free. Barker untied his legs, forced open the lid of the box, and crawled out. He didn't know if Wentzel was still in the house, so he quietly tried the door handle. It was locked. He moved to the window and tried to open it, but it was jammed tight. Then he thought he heard someone moving through the house. He quickly squeezed back into the box, retied his legs, fitted the broken cuff back to his wrist, and pulled the lid down over himself.

On the following night, 3 December, Wentzel pulled Barker out of the box to question him, and then left him on the floor while he went out of the room. Barker worked the handcuffs free again, pulled the hood from his head, untied his legs, and began to try to force the jammed window. Eventually, the window popped open with a loud bang. Barker was sure Wentzel must have heard the noise, so he sat down and pulled the hood back over his head. When Wentzel didn't appear, Barker removed the hood and pulled the window wide open. He squeezed through into a side alley, tiptoed into the back yard, climbed a fence, and ran for his life. Barker was free for the first time in eight days.

After running for two miles through the dark North Hollywood streets, Barker came across a gas station. He found a payphone, picked up the receiver, and made a collect call to his brother. Barker asked him to check his son was safe, and then call the FBI. Barker then called Kinnie Gibson, who told him to keep moving and get out of the area. A passing good Samaritan stopped to ask Barker if he was okay, and then offered him a ride to a nearby diner. Barker limped into the diner and called the FBI. They were waiting for his call, and told him to stay put. He watched nervously through the window, praying Wentzel wouldn't come after him. Within minutes, a car pulled up and two men entered the diner.

'Mr Barker?' asked one of the FBI agents.

'Yes,' said Barker. And he knew he was safe.

Barker was malnourished and dehydrated. He had lost almost 30 pounds in weight during his ordeal. His hands were swollen and numb, and his wrists were cut and sore from the handcuffs. He related to the FBI agents the bizarre tale of the RB-2000, and explained he had been lured to Los Angeles and kidnapped by men attempting to locate the rocketbelt. The agents drove him back to North Hollywood, and called in backup as Barker attempted to direct them to Wentzel's house. Then, about a block from the bungalow, Barker spotted Wentzel's car in the rear view mirror.

'That's him!' Barker said.

Within seconds, Wentzel's car was surrounded by unmarked FBI cars. This time it was Wentzel who found himself looking down the barrel of a gun. Or, more accurately, looking down the barrel of several guns. Wentzel was dragged from his vehicle, placed face down on the ground, handcuffed, and taken

154

away. His car contained two guns, and various tapes and ties in what was described as a 'death kit'. If he had been looking for the escaped Barker, it was fortunate he hadn't found him.

But for Barker the ordeal was finally over. The FBI agents took him to hospital, where he was treated for his injuries, and the next day drove him to the airport. The agents procured him a plane ticket, and Brad Barker went home.

THIRTEEN
THE LIFE SENTENCE

Brad Barker and Larry Stanley next faced each other on 25 April 2000, in Van Nuys Superior Court in Los Angeles. It was the preliminary hearing in the case of the State of California versus Thomas Laurence Stanley. Stanley was charged with kidnapping for ransom, false imprisonment by violence, and extortion. Christopher James Wentzel faced the same charges. Wentzel's preliminary hearing had been held in the previous month, on 6 March

Stanley had been charged two months previously, after turning himself in to the Harris County Sheriff's Department. He was then extradited to Los Angeles, where he was jailed, with a bail bond set at

$500,000. That was eventually reduced to $50,000, which Stanley paid, with some difficulty as his family's wealth seemed to be rapidly dwindling. He returned to Houston and attempted to raise enough money to pay for a defence attorney. He also fought extradition, as Los Angeles prosecutors requested a warrant for his re-arrest. Despite his best efforts, Stanley was eventually taken back into custody, and was forced to prepare to face his former partner in the courthouse.

Brad Barker sat on a wall outside the glass-fronted courthouse looking nervous as the hour of the hearing approached. He was wearing a dark suit and a red tie, and he wrung his hands and gazed at the ground. He was now 45 years old, and his once dark brown hair was salt and pepper grey. The events of the last ten years had visibly taken their toll. He got to his feet, lost in thought, and walked impassively past a bunch of reporters and a TV camera crew and into the courthouse.

The preliminary hearing was an opportunity for the court to test the charges against Stanley. The prosecutors would present evidence and supportive witnesses to the court to suggest probable cause to believe that Stanley had committed the offences he was charged with. The defence attorneys would then present contrary evidence and cross-examine the prosecution's witnesses in an attempt to rule out probable cause. Then Superior Court Judge Paul I

Metzler would decide whether the case would be sent to trial, or dismissed.

Stanley was led into court in handcuffs. The courtroom was a modern office-like room, with visitors and press sitting opposite the Judge's bench. Stanley, like Barker, had aged dramatically since the beginning of the RB-2000 project. He was now 55, his hair and moustache were grey, and his wrinkled face sagged in a befuddled frown. He was seated facing the Judge, and proceedings began.

First, Brad Barker took the stand. He told Deputy District Attorney Peter Korn that Stanley and Wentzel had lured him to Los Angeles with a job offer, then kidnapped him and placed him in a box for eight days. He outlined the violent nature of his ordeal, and told how he was deprived of food and water. Barker said he had suffered physically and mentally. Then he described the rocketbelt, and said Stanley and Wentzel had tried to ascertain its whereabouts. Barker said the men had threatened to harm him and his son if he did not comply.

Then Stanley's defence attorney, Leslie Abramson, questioned Barker. Abramson suggested that Barker could not be trusted, citing the hammer attack on Stanley, and the theft of the rocketbelt, as substantiation. Abramson said Barker had colluded with Joe Wright in an effort to cheat Stanley out of the rocketbelt's profits. On the whereabouts of the RB-2000, Barker said that Wright had held it for about a year after the partnership with Stanley dissolved. When asked where the belt was now, Barker claimed he didn't know. Abramson implied that Barker was lying, and said the kidnapping claim was a fabrication. In fact, said Abramson, Barker had been legally detained by men acting as bounty hunters, regarding the commercial burglary charge in Fort Smith, Arkansas. Abramson loudly chastised Barker for being unclear about certain details of the kidnapping.

'Like I said, I didn't sleep for eight days,' Barker explained. 'I was pretty in and out.'

Judge Metzler stepped in, telling Abramson, 'If you were in a box for eight days you'd probably have the same problem.'

'Yes, if I was,' replied Abramson, 'but I don't believe *he* was.'

'But if he was,' said the Judge, 'he doesn't deserve to be yelled at. Please modulate your voice.'

At the end of the preliminary hearing, Judge Metzler decided that there was probable cause to suggest that Stanley had committed the offences he was charged with. He ordered that the case go to trial before a jury. Wentzel's hearing had generated the same result. Both Stanley and Wentzel were released on bail pending trial. The case would not come to trial for almost two years.

Brad Barker was due back in court before that, as defendant in the commercial burglary charge relating to the incident at the Williams Tool shed in Fort Smith in 1999. But, on 12 February 2001, the District Attorney in the Greenwood District of Sebastian County, Arkansas, filed a nolle prosequi (an entry made into the court records declaring that a case will not be prosecuted). Essentially, the charges were withdrawn, but not dismissed. Although Barker was a free man, he could still be indicted on the same charges at a future date.

The reason for the non-prosecution of the case was never put on record, but it was suggested that one of the District Attorney's key witnesses – Vinson Williams – had refused to testify against Barker. The DA said Barker had threatened Williams and followed him. It was alleged that Barker had taken to turning up outside Williams' house in the middle of the night. Barker denied these claims, and said Williams had refused to testify for fear of being implicated in the kidnapping plot. Whatever the truth, Barker was a free man. The RB-2000, meanwhile, remained out of sight.

Stanley, no doubt wary of the fact that he was on bail, seemed to restrain his attempts to find the rocketbelt. He concentrated on pursuing legal avenues, and hired a new attorney to help build up a defence against the kidnapping charges. Both Barker and Stanley managed to keep low profiles in the run-up to the trial. Then the RB-2000 caper took another extraordinary turn.

On 27 March 2002, 28 months after the incident had occurred, the kidnapping case finally came to trial in Van Nuys Superior Court. Larry Stanley and Chris Wentzel were present as co-defendants, and they faced Brad Barker as chief witness for the prosecution. The press gallery was packed. Pool reporters had given coverage to the RB-2000 civil case and resulting $10.2 million judgment, and syndicated reports had appeared in newspapers across the world. But the bizarre kidnapping had really pushed the saga into the public eye, with big guns like the *Los Angeles Times* now keen to report on the resulting trial.

In his opening statement, Deputy DA Peter Korn explained how both Barker, now 47, and Stanley, now 57, were obsessed with the RB-2000, 'a backpack-like device that can lift a person into the air, if just for a few seconds.' Korn outlined the allegations, describing how Wentzel, now 54, had enticed Barker to Los Angeles. He said Barker arrived at Wentzel's home on 26 November 1999.

'The next thing he knows, he's in a chokehold with a gun to his head,' said Korn.

The Deputy DA went on to describe how Barker had been continually interrogated and threatened, and forced to sign an affidavit waiving his right to the rocketbelt. Korn said Barker was held for eight days, before escaping on 3 December. He also outlined Stanley's involvement in the kidnapping. Korn said he would show the jury email prompts sent by Stanley to Wentzel containing questions for Barker.

Stanley was now represented by defence attorney Dale Atherton, and Wentzel by Donald J Calabria. Atherton said Barker and Stanley decided to build the rocketbelt together, and invested $94,000 each in the project, but fell out in 1994.

'Mr Barker did not want Mr Stanley to fly the rocketbelt,' said Atherton, adding that the pair fought over padded manufacturing costs.

Atherton told the court Barker had attacked Stanley with a hammer, and stolen the rocketbelt. Despite Stanley winning a $10.2 million settlement granting ownership of the belt to him, Atherton explained, the belt had not been seen since 1995.

Atherton told the jury that the defendants had been acting as agents of the J&J Bonding bail bonds company in detaining Barker, as he had broken his bond by leaving the state for California. He said their action couldn't be construed as kidnapping, because Barker had come to California and Wentzel's house willingly.

Then Brad Barker took the stand. He recounted the details of the kidnapping ordeal.

'It's the most scared I've ever been in my life,' he said.

He explained how he had been lured to Los Angeles, and kidnapped by Wentzel. He said Wentzel had repeatedly questioned him, and threatened him with a gun, over the whereabouts of the rocketbelt.

'I thought he was going to murder me,' said Barker.

He then described how Wentzel and Stanley had told him, at gunpoint, to sign the affidavit.

'I did exactly as I was told,' Barker said.

The notary, Elyse Hoyte, was next to take the stand. She confirmed Barker had been handcuffed to a chair when he signed the affidavit. Then a J&J Bonding employee testified that she had authorised Stanley and Wentzel to detain Barker, under the false assumption that they were acting as agents of a licensed bail bondsman.

Detective John Krulac of the Los Angeles Police Department testified that there was evidence of correspondence between Wentzel and J&J Bonding. Krulac said it was his opinion Wentzel had believed he was holding Barker lawfully, but that had not been the case. He confirmed Wentzel had not been licensed to hold Barker, and said evidence collected from Wentzel's house supported Barker's testimony.

Sebastian County Sheriff's Department Detective Bill Hollenbeck confirmed in his testimony that the commercial burglary charges that had precipitated the bond had since been withdrawn. He said he believed Barker had been lured to California.

Then the court heard the testimony of Nancy Wright, Joe Wright's sister. She said Stanley had been convinced Barker had murdered her brother, and he had said to her that he would kidnap Barker, torture him, and kill him, if the authorities did not arrest him. Bill Suitor, who was also asked to testify, refused.

Throughout the trial, it became clear that both Barker and Stanley had different agendas to those of the defence attorneys and public prosecutors. Stanley was keen to use the trial as a platform to help him find the rocketbelt. And Barker used every opportunity during the trial to try to exonerate himself of the murder of Joe Wright. Their hidden agendas, eccentricities, and foibles – and their obsession with the rocketbelt – became clear to all present. But this court was not in session to find the rocketbelt, or to solve Joe Wright's murder.

After the evidence had been heard, the jury had the task of deciding whether Stanley and Wentzel had kidnapped Barker, or simply detained him as agents of the bail bond company.

On 25 April 2002, the jury delivered its verdict. They found Thomas Laurence Stanley and Christopher James Wentzel guilty on all counts of kidnapping for ransom, false imprisonment by violence, and extortion.

Stanley and Wentzel were led from the court by a female corrections officer. Sentencing was scheduled for 30 May. Both faced the possibility of spending the rest of their lives in prison. The jury had decided that, despite all of the twists and turns in the rocketbelt case, and the shady track record of the victim, Stanley and Wentzel had unlawfully kidnapped Barker.

In a fashion that had become typical in the many court cases involving the rocketbelt, the sentencing hearing was postponed for several months. It was 22 November 2002 when Larry Stanley and Chris Wentzel finally got to learn their fate.

Stanley and Wentzel were led into the court handcuffed together wearing blue v-neck prison overalls. Stanley shuffled into the whitewashed room looking slightly confused, almost as if he couldn't understand why he should be here. His hair was greyer than ever and his face was pale, with an apprehensive frown. This was the moment of truth. Wentzel looked less worried. He winked at a friend in the public gallery as he and Stanley were seated behind a wooden bench for the proceedings.

Wentzel was first up. He had entered into a plea bargain with the District Attorneys. He admitted to the charges of false imprisonment and extortion, in collusion with Stanley, but requested that the charge of kidnapping for ransom be dropped. Peter Korn agreed, saying Wentzel had been Stanley's lackey in the kidnapping.

'The worst thing that Mr Wentzel did was put Mr Barker in a box,' said Korn. 'That was an inhumane and despicable act.'

Wentzel's attorney, Donald Calabria, asked the

court to look kindly upon his client. 'This is a different man than many of the people you've sentenced,' Calabria told the court. 'If ever there was a man that would never be back here again under any circumstances, this is the man.'

Taking the plea bargain into account, Judge Barry A Taylor sentenced Wentzel to six years in state prison.

In contrast, Larry Stanley refused to admit his guilt. He had refused two offers of plea bargains from the DA, which would have meant accepting a three-year prison sentence. It was clear to the DA that Stanley was still obsessed with the rocketbelt, and would continue to pursue it upon his release. As a result, Peter Korn demanded Stanley receive a lengthy sentence, fearing he would re-offend if released within a short period of time.

Stanley simply refused to accept he had done anything wrong, and couldn't have expected the gravity of what was to happen next. He had sacked his attorney after the guilty verdict, and represented himself at the sentence hearing. Stanley sat alone at his counsel table while the sentence was delivered.

The courtroom fell silent as Judge Taylor explained how he had considered the evidence brought before him. Then the Judge revealed his decision. And it was a huge decision: Larry Stanley was sentenced to a state prison term of life plus ten years.

Stanley slumped over in his chair and placed his head on the table. It was the strongest possible punishment he could have expected to receive. His eyes welled with tears as he made a statement to the court.

'Your Honour,' he said. 'I never imagined that I ever did anything wrong. I was just trying to be persuasive. Mr Barker was not harmed in any way. I don't understand how this is a life sentence.' Eyes full of tears, he continued, 'My search for the rocketbelt has cost me more than half a million dollars and left my family destitute and on food stamps.'

Although not present in court, Brad Barker later declared himself to be pleased with Stanley's sentence, but unhappy with that given to Wentzel. 'I thought Wentzel's was really light,' he said. 'Larry Stanley, I thought he got what he deserved.'

Stanley was driven 400 miles north of Los Angeles to the California State Prison at Solano, a 146-acre medium security facility designed to hold 2,500 prisoners, but actually housing over 6,000. Two days later he sat down to write a letter on prison notepaper to Deputy DA Peter Korn. In the letter, for the first time, Stanley admitted his guilt.

'I owe you and the people of California an apology for taking so long to recognise my problem,' he wrote. 'My persistence and determination to obtain justice had become the obsession you spoke of in court on Friday.'

Korn had never believed Stanley's crime deserved a life sentence, but he had been frustrated by Stanley's refusal to accept his guilt. Now, with that situation changed, Korn decided to go back to Judge Taylor and ask him to reduce Stanley's sentence. It was an unusual move, but, under Californian state law, sentences can be modified within 120 days of being issued.

Korn returned to the Van Nuys Superior Court on Monday 13 January 2003. He told Judge Taylor he was confident Stanley would not re-offend upon release now that he had admitted his guilt. He asked the Judge to dismiss the charge of kidnapping for ransom.

'My job is to do the right thing,' he said, 'and I think this is the right thing.'

Judge Taylor agreed. He threw out the kidnapping for ransom charge and quashed Stanley's life sentence, but the charges of false imprisonment and extortion were upheld.

The Judge reduced Stanley's sentence to eight years in state prison.

Brad Barker also claimed to have had some say in the reduced sentence. After receiving a phone call from a mutual friend who told him of the distress of Stanley's wife and children, Barker said he contacted the deputy DA.

'I called Peter Korn and asked him to drop the

kidnapping charge,' said Barker. 'I said, "I'm not asking you to do this for Larry Stanley. I'm asking you to do this for his two children."'

Whether or not Barker's generous appeal had any bearing upon the decision, Stanley was a very relieved man. He now knew he would not have to spend the rest of his life in prison, although he would still have to endure incarceration for up to eight long years.

In the meantime, Brad Barker was free to return to a life of relative normality. He took a job as an electrician, hooking up instruments in Progen power plants. 'It's boring as hell,' he said. 'I hate it. I'd rather be building rocketbelts.'

And Barker hadn't forgotten about the RB-2000. Anything but. In fact, Barker was preparing with his attorney for yet another court case. He had filed a civil suit for assault and personal injury relating to the kidnapping, and the trial was set for June 2004. This time Barker was set to sue Larry Stanley, determined to win damages and have the 1999 civil judgment reversed. Aside from attempting to clear the multi-million dollar judgment against him, Barker was fighting for legal ownership of the RB-2000. And Barker was confident of a positive outcome.

'Hopefully,' he said, 'since I will be in court this time and Stanley will be in prison, the truth will be heard and things will maybe go a little better.'

That wasn't the case. Things weren't about to go even a little better for Brad Barker. After imprisonment, kidnapping, and accusations of murder, things were about to get even worse.

FOURTEEN
THE FINAL VERDICT

For Brad Barker, his legal challenge seemed a final throw of the dice. He had spent almost four years and tens of thousands of dollars building his precious rocketbelt. It had worked, but he had hidden the device away after falling out with Stanley and Wright. Then Wright had turned up dead, and fingers had been pointed at Barker. A Judge subsequently awarded legal ownership of the rocketbelt to Stanley, but Barker refused to hand it over. So Stanley had him kidnapped. And the worst of it was, after all that had happened, Barker was still unable to fly the rocketbelt. Even with Stanley in prison, Stanley's family and attorneys were still searching for the belt. It had to remain hidden.

Legally, Barker couldn't even admit to knowing its whereabouts. So, when asked if he knew where the Rocketbelt 2000 was, Barker repeatedly claimed ignorance.

In fact, Barker knew exactly where the rocketbelt was. Since the break up of the partnership, the belt had been moved between various locations for safekeeping. Barker himself said during the kidnapping trial that Wright had held onto the rocketbelt for about a year after the break-up of the partnership with Stanley. Shortly before his death in 1998, Wright had told Stanley that Barker had taken possession of the belt, and had stored it in North Harris County. The northernmost town in Harris County is Tomball, the place where Barker held a lease on a storage facility. But by the time Larry Stanley located the storage facility in November 1999, the rocketbelt had gone.

By that time, Barker was living in Fort Smith, working on Vinson Williams' flying can. According to Tom Wade, the RB-2000 was stored there, and the men played around with it. Stanley had continued to follow the trail of the rocketbelt, driving around Texas and Arkansas trying to catch up with Barker, but he was always one step behind. Barker had moved the belt to Dallas, then back to Houston, and then to a storage facility in La Porte, 30 miles east of Houston. During his kidnapping ordeal, Barker told his captors that the rocketbelt was being held by a friend as collateral in a loan deal. There was even talk of Barker burying the rocketbelt in a friend's backyard.

Eventually, Barker passed the rocketbelt to Rob Fisher, the karate instructor who had helped him out with the dispute at Stanley's oil field. But Fisher became wary of keeping the belt on his property. He had become nervous since hearing the details of Barker's kidnapping, and he didn't want to get dragged into a similarly dangerous situation. So Barker and Fisher came up with a plan. The rocketbelt would be broken down into pieces and vacuum-sealed inside the specially-designed watertight containers. The containers would then be submerged underwater, at one of the many nearby rivers and reservoirs, or perhaps buried at a local beach. The belt would then be entirely safe from detection by the Stanley family, and Barker could truthfully state that neither he nor his friends had possession of the rocketbelt. Once the court case had been settled in Barker's favour, the parts could be reclaimed.

The plan was eccentric and didn't make much sense, and so was perfectly suited to be associated with the Rocketbelt 2000. In any case, Barker held off from sinking his precious rocketbelt. His assault and personal injury civil suit was coming to court, and he was very confident of a positive outcome. Barker expected to have the 1999 civil judgment overturned and be awarded legal ownership of the belt, plus receive damages from Stanley. He was wrong.

On 29 June 2004, Judge Martha Hill Jamison of the 164th Civil District Court in Houston dismissed Barker's case for want of evidence. Key to Barker's argument was the claim he had suffered damaging mental trauma during the kidnapping. However, Barker and his attorney failed to provide any corroborating evidence to that end. Normally in such circumstances a complainant would seek consultation with and testimony from a psychiatrist. Such consultation and testimony were not sought. As if the dismissal wasn't bad enough, the Judge also charged Barker with contempt of court. Barker had never responded to the court's award of $10.2 million to Stanley in 1999 and, specifically, he had never surrendered the RB-2000 as required. Barker was ordered to appear on the contempt charge in front of Judge John T Wooldridge, the original Judge from the 1999 rocketbelt trial.

Following the mistake over the evidence in the civil case, Barker dispensed with the services of his attorney, but he was not able to replace him. So Barker duly appeared at the 269th District Court on 12 July 2004, but brought no counsel with him. Judge Wooldridge ordered Barker to appoint a new attorney, and postponed the hearing by a week.

'The Judge was very kind,' said Barker after the

court appearance. 'I've got to go speak to an attorney. But the Judge was very nice to me, and actually kind of short to the other side it seemed like, so we'll see what happens.'

One week later, on 19 July, Barker did turn up with a new attorney, but claimed he had not had sufficient time to prepare. Judge Wooldridge gave Barker a continuance order, and the case was reset for 13 August. The Stanley family was present at the trial dates, and was irritated by Barker's hindrances. So was Judge Wooldridge. Perhaps recalling the delays that had beset the 1999 trial, the Judge gave Barker a stern warning. He ordered Barker to produce the Rocketbelt 2000 in court, or face the consequences.

On Friday 13 August, Barker and his attorney turned up in court with a large metal box, so heavy that it took several men to drag it into the court room. Anticipation was high. Those present craned their neck for a view of the contents, hoping they were about to get a glimpse of something that had not been seen in public for nine years. Had Barker considered the Judge's warning and finally handed over the rocketbelt? No. The box was opened to reveal not the rocketbelt but a pile of spare parts.

What Barker was trying to achieve by this frustrating move was unclear. If handing over a box of spare parts was intended as some sort of conciliatory act, then it failed. It only succeeded in angering the Judge and the Stanley family. When questioned, Barker offered the explanation that he was having difficulty getting hold of the RB-2000. The judge was unimpressed, but nevertheless set a new court date for 24 September, again ordering Barker to present the rocketbelt in court. He also put Barker and the Stanley family into mediation.

Nothing that is said in court-ordered mediation can be used in trial. So, Barker was free to admit to the Stanley family that he had possession of the rocketbelt, something that they already believed to be the case. Therefore the discussion concerned not whether Barker had the rocketbelt, but whether he would hand it over. Mediation discussions between Barker, the Stanley family, and their attorneys lasted over six hours, and initial progress was made. The Stanley family offered to release Barker from the $10.2 million judgment, and give him what they considered to be a fair percentage of the future profits if he handed over the RB-2000 in flight-worthy condition with all related parts and components. But Barker refused. He said he would consider nothing less than a full 50 percent share in the rocketbelt. The Stanley family's counsel dismissed this, and closed the mediation.

After all that had gone on, and with Larry Stanley in prison, the Stanleys were understandably sick and tired of the Rocketbelt 2000. Clearly, they were prepared to go to great lengths to ensure that

Barker would not keep hold of the rocketbelt. But were they still keen to get their hands on the troublesome device? Larry Stanley's attorney, Michael Von Blon, said his client would pursue entertainment bookings for the belt in order to recoup his tens of thousands of dollars worth of investment into the project. But the Stanley family had other ideas. They declared that, once they reclaimed the belt, they would offer it up for sale because of the painful memories it held for them.

In truth, Barker couldn't have spent much time considering the Stanleys' offer. In September 2004, just weeks after the mediation broke down, he headed to Rob Fisher's house and reclaimed the rocketbelt. If he was planning to go ahead with the plan to submerge the belt, he didn't let on to Fisher. He simply loaded up the belt into his truck and drove off, just as he had after the public flight in 1995. Whether or not he dumped his prized rocket in one of Houston's many waterways remained unknown.

The court date of 24 September was rescheduled by the Judge for unspecified reasons, and instead the case resumed on the afternoon of 8 October. Barker and the Stanleys were present in court on that date, but the Rocketbelt 2000 was again conspicuous by its absence. Barker maintained that he did not know where it was. An exasperated Judge Wooldridge had now had enough. Larry Stanley's civil suit had been filed in 1995, nine years previously. Over those years, Judge Wooldridge had become very familiar with Brad Barker's face, and had regularly been forced to reschedule court dates due to Barker's delaying tactics. No more.

Judge Wooldridge found Barker guilty on three charges of contempt of court. He then gave Barker one full week, until noon on 15 October, to present before him the Rocketbelt 2000 and all associated components. And he offered Barker a stark ultimatum. Should he not surrender the rocketbelt as ordered, he would be fined $500 and immediately taken into custody. Judge Wooldridge told Barker he would serve six months in jail for each charge, a total of 18 months behind bars, and this sentence could be extended indefinitely until he handed over the rocketbelt.

Now Barker was in a real bind. He stated once more that he did not have the rocketbelt. But he had been seen with the device just weeks previously, when he had picked it up from Rob Fisher's place. Indeed, Barker's attorney said his client would honour the Judge's request, and therefore must have believed that Barker did have the rocketbelt. And even if Barker did not physically have possession of the device, it was most likely that he knew where it was hidden. It was quite possible that the belt was stashed either at the home of another friend or family member. Even if Barker had gone through with the plan to submerge the belt under-water, he could still go about retrieving it. Barker had a simple choice: give up the RB-2000, or go to jail. He had already spent several months behind bars in relation to the rocketbelt caper, and had hated being separated from his son. The decision seemed obvious.

The Stanley family seemed to think so. Of course, they wanted Barker to give up the rocketbelt, but they also expressed sympathy for Barker's son, and said they did not wish to see Barker incarcerated. This was despite the fact that the feud between Barker and Larry Stanley over the rocketbelt had seen Stanley jailed. They were cautiously optimistic that Barker would hand over the RB-2000.

Yet Barker remained defiant. Judge Wooldridge had ordered that he hand over the rocketbelt and its associated components, including the fuel laboratory trailer. But Barker did not have the trailer. That had been taken by Larry Stanley when he came across it in Fort Smith during his search for the rocketbelt five years earlier. Barker claimed that the Stanley family would be aware of that, and therefore it was an impossible demand. He planned to offer this fact to Judge Wooldridge, and was confident this would see the contempt charges dismissed.

So, Barker turned up in court on Friday 15 Octo-ber empty-handed, and offered up only excuses.

Judge Wooldridge was entirely unimpressed. He called Barker's refusal to surrender the rocketbelt, 'a blatant attempt to subvert this court's jurisdiction and enforcement of its lawful orders.' Then Judge Wooldridge signed an order for Barker to be held in the Harris County Jail until such a time as he decided to surrender the rocketbelt.

Barker, now 50 years of age, was immediately taken to the nearby jail and locked away. What was going through his mind could only be speculated upon. Although he made confident noises ahead of the court appearance, he must have known that, if he refused to give up the Rocketbelt 2000, he was going to end up behind bars. Yet he decided not to surrender the belt. He would rather go to jail indefinitely than hand it over, such was his love for the rocketbelt, and his hatred for Larry Stanley.

If Barker and Stanley could see clear blue sky from the windows of their respective jail cells, then maybe they would still dream of flying free by the power of a rocketbelt. Joe Wright, buried in Michigan, could no longer dream such dreams. And the RB-2000, the amazing machine that had driven these men to distraction and beyond, was hidden away, perhaps buried or submerged under water. It had flown only once in public. Where was it, and would it ever fly again? Only Brad Barker knew the answer to that question, and, sitting alone in his Houston jail cell, Brad Barker wasn't answering.

FIFTEEN
THE AFTERMATH

Fifteen years after Brad Barker, Larry Stanley and Joe Wright first set out to build their rocketbelt all that was left were ruined lives and unanswered questions. It seemed likely that the truth about the murder of Wright and the location of the rocketbelt would never be known.

Larry Stanley's civil attorney, Michael Von Blon, reflected upon the caper: 'One man is dead, one man is in prison – whether or not he is guilty I don't know – and the other man, who has never done anything right in any of the chapters of this story, is running free.'

Doug Malewicki worked closely with Barker, Stanley, and Wright on the rocketbelt project. 'Larry

Stanley let his hate for Barker get the better of him and he ended up with an eight-year prison term,' he said. 'He should have just let it go, and got on with his life. Too bad.'

The most distressing aspect of the story remained the death of Joe Wright. 'I will remember Joe for the rest of my life,' said Wright's close friend Maurice Heimlich. 'He was a wonderful person. He would do anything to help out people that he knew and cared for. We were great friends and I still can't believe that he is gone.'

Bill Suitor knew Barker, Stanley, and Wright very well, having worked and lived with them during the building of the RB-2000. He became particularly close friends with Wright. 'Joe Wright was one of the kindest, most generous people I ever knew,' said Suitor. 'I cannot think of one bad thing to say about him. Nor one good thing to say about Brad Barker. As for Larry Stanley – what a loser. Barker and Stanley were obsessed and blinded by the notion they were both going to learn how to fly and become rich and famous. And that's all they cared about – being rich and famous. So much so that now Joe is dead. And none of them is rich or famous. If anything, they are infamous. As for me, it was the dumbest thing I ever got involved in. The dumbest. I do not normally grant interviews on the subject, and that will be all I will ever tell you about that sad chapter.'

Nancy Wright and her family were left to pick up their lives without a beloved brother and son. They remained hopeful, but not optimistic, that the truth about his death might one day be revealed, and justice might be done. In the meantime, they watched what little money Joe Wright had left disappear at the hands of his creditors. 'There was nothing left in Joe's estate,' said Nancy Wright. 'He had lost everything. The bank took his house and his car. But they never went after Brad. He is hard to track down as he has never had a real address or a real job. I think that is part of the reason Larry had him kidnapped. Larry had tried to go after Brad legally and won, but nothing happened, so he took matters into his own hands.'

So who did kill Joe Wright? Although Brad Barker remained a suspect, there was no evidence to link him to the murder. He had been evasive, and his alibi did not quite fit, but that did not necessarily make him a murderer. Larry Stanley had no apparent motive, but had revealed disturbing characteristics during the kidnapping debacle. Again, however, there was no evidence to link him to the crime. But the timing of the murder – just hours after the speakerphone meeting with Stanley, and days before the scheduled start of the civil trial – strongly suggested a link with the Rocketbelt 2000. Could Barker or Stanley have hired another individual to kill Wright? Or did the murder have nothing whatsoever to do with the RB-2000 after all?

Wright's friends had also named bookmaker Mike Bowman and drug dealer Wayne Johnson as suspects. Wright and Bowman had fallen out over money, but, as a bookmaker, Bowman must have been used to that sort of conflict. It seemed unlikely that the money wrangle alone would have led to murder. Wright had also fallen out with Bowman's employee Johnson over money. Again, however, as an employee of a bookmaker, Johnson must have been used to dealing with non-paying clients. Killing Wright would not have helped in recovering the debt, but Johnson and Wright were involved in drugs, and this murky factor could have precipitated a deadly situation. Or perhaps Bowman and Johnson colluded over the killing. When Johnson was unable to recover the debt, did Bowman order his employee to kill Wright?

The gambling ring angle remained equally unclear. It was alleged that Bowman's illegal gambling ring involved members of the Harris County Sheriff's Department. Wright, a friend of both Bowman and a Lieutenant in the Sheriff's Department, had allegedly drawn up a file containing information about the gambling ring. The existence of such a file could, in theory, have provided Bowman with a motive to kill, or order the killing of Wright, and encouraged the Sheriff's Department to bury the investigation. The file supposedly disappeared while in the custody of the Sheriff's Department. But the FBI were investigating Bowman and Johnson as part of an investigation into illegal gambling. Surely they would have uncovered any such gambling ring, and related conspiracy?

So little was known about the suspect known only as 'Diane' as to be useless to the investigation. It was said that Wright had a dinner appointment with Diane scheduled around the time he was murdered. But it seemed only Wright knew who Diane was. Wright had been dressed to go out when he was killed. Why did Diane not come forward after Wright's murder? Did she have something to do with it? Was she the killer, or was she used by the killer to persuade Wright to open the door? Or did she have something else to hide? It was suggested that she might have been Wright's crystal meth dealer. And with Wright leading such a secretive life, did Diane really exist at all?

There were also suspicions over Wright's former partner, Bryan Galton. Galton did have a history of violence against Wright, and had recently returned to the Houston area. But he could not be questioned with regard to the murder. Galton apparently died from a massive drug overdose not long after Wright was murdered. It was alleged that the overdose was a suicide, driven by guilt. Galton had a close per-sonal relationship with Wright, and the manner of the murder suggested a personal killing or crime of passion. Could the fractious relationship between Galton and Wright have led to murder? Or could the murderer have been another boyfriend, propelled somehow into a violent rage?

The Wright family continued to push the Sheriff's Department and the District Attorney's Office to reinvestigate the case. Then Wayne Johnson was arrested on charges of driving under the influence and assault with intent to murder. Johnson had been involved in a car accident while drunk, and had beaten up his elderly father when he came to offer assistance. Johnson, 49 at the time of his arrest, had a long criminal record that ran back more than 20 years and included assaults, drug offences, driving offences, and other misdemeanours. He also drove a black Toyota Avalon, which matched the description of a car seen outside Joe Wright's house on the night of the murder.

While in custody, Johnson was questioned for the first time about Wright's murder. Detectives said he appeared nervous, and refused to take a polygraph test on the advice of his attorney. Johnson was bailed and secreted into rehab before the investigation could continue. The assault charges were dropped. His car was never searched. Another trail ran cold, and the murder of Joe Wright remained unsolved.

Brad Barker was into his sixth month in Harris County Jail when the call went out: *'Barker ATW.'* Sitting in his jail cell, Barker ignored the call. 'It just went in one ear and out the other,' he said. Under the terms of his incarceration he could be held indefinitely, and Barker was resigned to a long stay. 'I just wasn't expecting to get out,' he said. 'On contempt they can keep you in forever. And I'd really pissed that Judge off.' But 'ATW' meant 'All The Way'. It meant Barker was going home.

A guard peered into the cell and confirmed the news: 'Barker, get your shit. You're out of here.'

'I don't know why the Judge decided to let me go,' said Barker. 'After he put me in jail I wrote him a three-page handwritten letter basically telling him that what he did to me was a bunch of bullshit. That probably didn't help much, but I'd had enough of his shit. I told him the way this whole thing was handled was a joke. I told him the Harris County Sheriff's Department was as corrupt as hell. It probably wasn't a smart thing to do, but I was tired of being screwed with.'

It was April 2006 and, after five and a half months in jail, and despite his letter to the Judge and his continued refusal to surrender the Rocketbelt 2000, Brad Barker was a free man. He began to pick up the pieces of his life, spending time with his son, and returning to work, but the RB-2000 remained at the forefront of his mind. Barker was preparing for another round of court-ordered mediation, and this time he was seeking resolution.

Barker was now represented by attorney Jeff Haynes. Although Barker and Haynes hadn't been aware of the fact at the time, the two men's paths had crossed before. Back in 1999 when Barker was in the Sebastian County jail on commercial burglary changes, Haynes was incarcerated in the same facility. Haynes had been going through a hard time following the death of his wife. He'd quit practicing law, got involved in drugs and, in his own words, 'my life was basically out of control.' He eventually ended up in jail after being busted for drugs. As a criminal defence attorney he'd made some enemies in law enforcement, and Haynes claimed he'd been set up and forced to plead guilty to avoid trumped up charges being levelled at his sons. So Barker and Haynes were in the same jail at the same time, although they didn't properly meet until they were introduced several years later in Haynes' law office. 'I find this to be a rather strange coincidence,' said Haynes.

Barker and Haynes were determined to free the rocketbelt from its legal wrangles. Barker seemed willing to offer some concessions to the Stanleys in mediation, in the hope that the rocketbelt could one day fly again. Haynes thought it was entirely possible that the situation could be resolved, 'but due to the nature of litigation in this country you don't ever have a clue exactly what's going to happen.'

Barker said he was sick of the whole affair, and said he no longer dreamt of flying the rocketbelt. 'It's a nightmare,' he said. 'I just want it to end.' But, with mediation hearings being continually post-poned and reset, it didn't look like the nightmare was going to end any time soon.

Then Barker was waylaid by another turn of events, this time involving Stan Casad - the ac-quaintance who had helped lure Barker to Chris Wentzel's home in advance of the kidnapping affair.

Casad was in trouble. He had been providing tes-timonials for a company called HEE Corp regarding a supposed miracle cure for diabetes. In 2004, Casad and an associate named Ron Brooks had, it was claimed, travelled to the United Arab Emirates to undergo treatment for the condition. They had subsequently reported that their blood sugar levels had fallen 'dramatically to levels normally seen with non-sufferers'. Press release agencies and stock trading message boards were blitzed with news of Casad and Brooks' remarkable progress. HEE's profile soared, and its shares were labelled as hot tips.

But Brooks and Casad's testimonials were hardly impartial. Brooks was revealed to be a major share-holder in HEE, and in 2005 Casad was named as the company's new president. The pair spent com-pany proceeds on property, cars, a helicopter and a

Gulfstream II passenger jet. But the spending spree didn't last long. Brooks became embroiled in an investigation over the theft of $500,000 worth of HEE products, the company was hit by a civil judgement and unpaid tax warrants totalling almost $1 million, and the US Food and Drug Administration issued a warning against selling unapproved drugs. Then, in early 2006, the company's entire board resigned. As HEE Corp collapsed, investors were said to have lost around $40 million. Brooks and Casad were wanted men.

Casad needed to cover their tracks, and he called Brad Barker for help. When Barker refused, he says Casad threatened his life. Barker ended up driving Casad to the airport, believing he was flying off on a business trip. It turned out Casad was fleeing the country. He was thought to have ended up in Ukraine, telling friends he would never be coming back. Barker was less than happy. If Casad did ever return to the US, he would have to deal with both the authorities and a very angry Brad Barker.

It was yet another twist in an extraordinary tale. Barker admitted that his story was 'crazy' and had all the elements of a Hollywood movie, but who would he like to see play him onscreen? 'Pee Wee Herman!' Barker said. 'Or maybe Brad Pitt. You can't go wrong with Brad Pitt.'

Larry Stanley's sentence ran until 2010 and, with state law ruling that prisoners convicted of violent crimes must serve at least 85 percent of their term, he could not be released before December 2008. He had been moved from state prison to a medium security Correctional Training Facility in Soledad, and had subsequently applied to be transferred from California to a facility in Texas to be nearer his home. Stanley had expressed regret for his obsession with the rocketbelt in order to have his original life sentence reduced, but had he really put the affair behind him? When asked in a letter whether he would continue his search for the RB-2000 upon his release he declined to answer.

Joe Wright's murder, meanwhile, was no nearer to being solved. Perhaps the murderer was one of his rocketbelt partners, but there remained no evidence to link either Barker or Stanley to the crime. Or perhaps the murder had nothing whatsoever to do with the rocketbelt, and the timing of it had simply been a coincidence. There were plenty of other suspects, but insufficient evidence meant a conviction seemed unlikely.

And then there was the rocketbelt, the amazing invention around which the whole caper had revolved. It had not been seen in public since the demonstration flight at the Houston Ship Canal in June 1995. It seemed likely that Barker still had possession of the rocketbelt, or at least knew where it could be found. Was it hidden away, perhaps buried or submerged? Would it ever fly again? The whereabouts and future of the Rocketbelt 2000 remained a mystery.

SIXTEEN
THE NEW ROCKETEERS

The RB-2000 might have been grounded, but it was not the only rocketbelt in the world. Powerhouse, Kinnie Gibson's company, now owned three rocketbelts, including the original Tyler belt. They had built their two new belts from scratch, making improvements to the Bell and Tyler designs, and said that both could fly for more than 30 seconds. Powerhouse continued to make promotional flights around the world, although Gibson himself had retired from flying rocketbelts. Gibson continued to fly hot-air balloons, competing successfully in international balloon races, and continued to be involved in dramatic incidents.

In November 2003, Gibson was one of four

passengers in a twin-engine Cessna that crashed on landing at Fort Worth, Texas. The plane was struck by a strong gust of wind as it came in to land, and hit the ground 50 yards short of the runway, bursting into flames. The passengers survived the crash, but found themselves trapped inside the burning plane. They desperately struggled to open the pressurised cabin door. Then Gibson stepped forward, and kicked the door three times, bursting it open. He led the passengers to safety as, behind them, the plane became engulfed by a fireball of such intensity that the aircraft quickly melted into a puddle of molten metal and plastic on the ground. None of the passengers were injured, thanks to Kinnie Gibson. The former stunt double to Chuck Norris had become a hero in his own right.

Gibson continued to own Powerhouse and Rocketman Inc, but he handed over the flying duties to Eric Scott and Dan Schlund. Scott, from Montana, was an ex-US Air Force Paratrooper. He moved into professional stunt work after leaving the armed forces, and also worked in acting and special effects for the film and television industries. He held the record for the highest human flight with a rocketbelt, soaring to 152 feet in London in 2004. Schlund, from Valencia, California, was another stuntman, and a former emergency paramedic. Schlund estimated Powerhouse had performed around 1,200 flights since the mid-1980s. For him, there had been many highlights.

'Personally, I love seeing parts of the world that I would never see, and being part of events that I would never be part of,' he said. 'But also, it's the people I've met, worked with, and struggled with. And I've found that everyone, from Brazil to Japan to Saudi Arabia, has the exact same smile, and awe and excitement in their eyes. Highlights include the Rio de Janeiro Carnival, Michael Jackson's *Dangerous* tour, and flying for President George W Bush, to name but a few.'

Since the disappearance of the RB-2000, Powerhouse had owned the only commercial rocketbelts in the world. If you wanted to hire a rocketbelt, Powerhouse was your only option. Kinnie Gibson even outlined plans – subsequently foiled – to register the word 'rocketbelt'. His company held a monopoly on the lucrative rocketbelt demonstration market. But then a rival rocketbelt company emerged.

Troy Widgery was the founder of Go Fast Sports, a company that had supplied skydiving equipment to Kinnie Gibson. A former world-class skydiver himself, Widgery, from Denver, Colorado, had also been involved in a horrific plane crash – he was one of the few survivors of a tragic training flight crash that killed sixteen people in California in April 1992. Having made millions from his sports company, Widgery had launched a Go Fast energy drink, and in 2003 he decided that a rocketbelt would make a great promotional tool. He set up a company called Jet Pack International and spent three years and a million dollars building the Go Fast Rocketbelt. Then he poached Powerhouse pilot Eric Scott to fly it.

The Go Fast Rocketbelt was lighter than any of its predecessors, and it was claimed that it could fly for 33 seconds and to an altitude of 300 feet – longer and higher than any other rocketbelt. It looked pretty good, too, polished silver and red with Go Fast decals. The pilot was decked out in a black Go Fast jumpsuit and helmet. Powerhouse also smartened up their belt, dressing their pilot in a new futuristic *Robocop*-style outfit.

Both Powerhouse and Go Fast performed rocketbelt flights around the world, although both encountered problems. A Powerhouse flight in Australia ended with Dan Schlund crashing to the ground on his knees. And a Go Fast flight in Mexico disappointed thousands as Eric Scott and his rocketbelt failed to lift into the air. But there were triumphs as well as disappointments. Dan Schlund donned make-up and sunglasses to double as rap star P Diddy in a stunt for the 2005 MTV Awards, and his spectacular New Years Day 2007 flight at the Los Angeles Tournament of Roses parade was broadcast around the world to millions. Eric Scott flew at Go Fast 'Xdays' in Germany and the Netherlands in June 2006, and he also performed demonstration flights at the first International Rocketbelt Convention, at the Niagara Aerospace Museum in New York in September 2006.

The Rocketbelt Convention attracted enthusiasts from around the world – and saw Eric Scott meet up with fellow rocketbelt pilots Hal Graham, Bill Suitor, Nelson Tyler, Peter Kedzierski and John Spencer. As resounding success, plans were immediately prepared to make the International Rocketbelt Convention a yearly event. The convention was organised by Kathleen Lennon Clough, daughter of Bell cameraman and test-rig pilot Tom Lennon, along with Peter Gijsberts, proprietor of the exhaustive rocketbelt website www.rocketbelt.nl.

Gijsberts managed a corporate transport service in the Netherlands, but hours of research in his spare time made him a walking authority on the history of rocketbelts. 'I have a memory of seeing two guys flying rocketbelts over some woods from my childhood,' he said. 'For a long time I thought that might have been a dream. But then a couple of years ago I found out about the reality of rocketbelts, and I began to investigate and build a website.'

Gijsberts proceeded to telephone, fax and email scores of rocketbelt enthusiasts in an effort to find information and photographs for his site. He discovered that a small band of dedicated enthusiasts in workshops and garages across the world were attempting to build their own rocketbelts. All believed that one day they would fly through the air with a rocketbelt strapped to their back. And so they dedicated large sums of money and countless man-hours – and also risked severe injury in both the building and testing phases – in the hope of one day owning their own amazing rocketbelt.

Frank Dickman, a mechanical and aerospace engineer who worked as a consultant with Boeing and NASA, was one such rocketbelt enthusiast. 'As a university student, I had considered building a rocketbelt,' he said. 'Unfortunately, I had no money, no practical experience, and no industry supplier contacts. But in the intervening years after college I gained experience in all the industries required to design and build an improved version of the Bell Rocketbelt. After a week or two of research, I concluded the rocketbelt could possibly be built better, cheaper and faster than at the time of its invention.'

Dickman set about building his rocketbelt, with the trademarked name EagleOne, in his living room. He had uncovered a raft of information, including the original rocketbelt patents and construction plans, but found a lot of it to be incomplete and unreliable.

'It's a trail of the cryptic notes and sketches and unwritten concepts in the minds of long dead engineers,' Dickman said, 'and perhaps it is a trail littered with the souls of dreamers.'

Stuart Ross, a Boeing 767 airline pilot from Sur-
rey, England, began to build a rocketbelt with the
aim of flying it at paid bookings around the world.
He planned to offer the belt up for air shows, corporate events, and even as an extra special entertainment for birthday parties. Ross eventually planned
to build five rocketbelts and offer rides to the public.
Ideal for extreme sports fans who thought bungee
jumping or skydiving was too tame. But making
money was of secondary importance to Ross. Building his rocketbelt was a labour of love.

'I'd always dreamt of strapping on something like
this and just flying around,' said Ross. 'I've always
loved flying. I've clocked up ten thousand hours of
flying planes now, but a rocketbelt is completely
different. A lot of my colleagues at work get involved
in flying old restored fighter planes and things like
that, but I thought, sod it, let's go for something a
bit different. I've always wanted to do something
like this, so off I went with it.'

The testing of the rocketbelt was delayed to ac-
commodate Ross's wedding. His wife-to-be didn't
want him hobbling down the aisle in a body cast.
'She wouldn't have that,' he said. 'So I had to knock
the rocketbelt on the head for a couple of months.
But it would have been fun to turn up to the wedding on it!'

In preparation for flying his device, Ross sought
advice from the UK's Civil Aviation Authority

(CAA). 'The CAA were very interested in the rock-etbelt, and quite positive about it, but they thought I was a bit crazy,' said Ross. 'At the end of the conversation they said there was another chap who wanted a quick chat with me. It was a psychiatrist from their medical unit.'

Some of Ross's friends and colleagues also doubted his sanity. 'I get emails from guys at work telling me that the best place to test the rocketbelt is in the car park of the local hospital, because then I'll not have far to go if something goes wrong,' he said. But for all the difficulties and dangers Ross encountered during his building of the rocketbelt, one sacrifice hurt most of all. 'The hardest thing so far has been cutting out the beer,' he says. 'Losing the beer gut means you can carry more fuel, so you have to cut out the beer!'

Ross suffered peroxide burns and sore ears from the incredible noise generated by his device, and other rocketbelt projects proved no less problematic. Stuntman Tom Edelston designed, built and tether-tested a rocketbelt in 2001. The testing didn't always go to plan and, on one occasion, an overflow in the rocketbelt's engine caused H_2O_2 to spill from the exhausts and set fire to Edelston's leather boots. Then a fire broke out in Edelston's workshop, and his rocketbelt and associated equipment went up in flames.

Arnold Neracher, the 'Swiss Rocketman', also

built a rocketbelt. In fact, he built two. A rocket engine expert, Neracher also built a rocket bicycle, and a rocket go-kart called the Turbo-Wankel. In 2006 he made the amazing claim that one of his rocketbelts had flown for a full minute, almost double the flight duration of any previous belt. It was then reported that he was set to unveil a rocketbelt that would fly for six minutes. Other rocketbelt builders also offered astonishing claims.

Nino Amarena of Thunderbolt Aerosystems in Los Angeles claimed to have successfully built a new generation rocketbelt, called the Thunderpack, that could run for up to 90 seconds, or for 140 seconds in 'bi-propellant mode'. Amarena said the US Coast Guard had expressed an interest in purchasing the Thunderpack for ship-to-ship rescues, and he claimed to have already sold packs to the Far East for use in earthquake rescue. Amarena planned to have Thunderpacks available for purchase by the general public, and he recruited Bill Suitor to test the device in tethered flights.

But perhaps the most amazing claim of all came from Juan Manuel Lozano and his company Tecnologia Aeroespacial Mexicana. TAM, a Mexican developer of hydrogen peroxide rockets, began offering 'a complete turn-key package of a flying rocket belt, custom-made to the pilot's weight and body size.' Complete with a fuel lab and training, the price was $250,000. Lozano had been building rocketbelts for several years, and said his TAM design had been fully tested and proven. Other companies sold rocketbelt parts and plans via the internet, but Lozano offered a stark warning: 'Be aware of people that offer plans, parts, or a rocketbelt that has not flown and tested because you could be killed.' Lozano claimed to have flown one of his belts himself, and had built one for his daughter Isabel – the world's first female rocketbelt pilot.

The TAM Rocketbelt looked a tempting purchase. 'It could be a goldmine in the right hands,' said Lozano, who admitted that financing the project had pushed his bank account into the red. Unfortunately, sales figures for the world's first commercially available rocketbelt were not forthcoming.

Then Troy Widgery announced he would sell copies of his Go Fast Rocketbelt for $155,000. He also invited pre-orders for a brand new jetbelt, the $200,000 Jet Pack T-73, with a reported flight time of 19 minutes. But the T-73 was still in development and, when completed, would only be sold to 'qualified individuals'.

Despite commercial versions being made available for purchase, the rocketbelt remained too expensive and too limited in terms of flight time for general use. The idea of strapping on a rocketbelt to commute to work remained a fantasy. So when would the general public finally get an affordable, reliable, working rocketbelt?

'Probably not in ten years, but possibly in our lifetime,' said Frank Dickman. 'Theoretical physicists are working on a variety of new propulsion systems, most of them very exotic, and theoretical. If you are young enough, you may get to see them.'

'I personally don't think these gizmos will be widely available in the future,' said Peter Gijsberts. 'They'll only be used in the entertainment industry, due to high expense, high danger, and the difficulty of building, testing, and flying the things. And the flight times are too short. After 28 seconds or so, the thrill is gone.'

So could the rocketbelt design be improved to make the device more practical? In the many years since the invention of the Bell Rocketbelt not much had changed at all. The very latest rocketbelt designs showed no great technological leaps beyond those of the Bell device, and many actually seemed to be inferior to the original design.

'The limitations of the rocketbelt are a result of the laws of physics, which, unlike other laws, are difficult to evade,' said Frank Dickman. 'The thrust required to hover, in a rocketbelt or a Harrier jet, requires notoriously high fuel consumption in comparison to aerodynamic flight. The rocketbelt consumes fuel faster than a Boeing 747 passenger jet. You could improve the rocketbelt, at great cost, but the best result would only offer around sixty seconds of flight.'

So the rocketbelt is unlikely to ever fulfil its iconic expectations, regardless of technological advances. The unsuitability of the rocket engine means the rocketbelt is fundamentally flawed. What about alternatives?

'I think jetbelts have a brighter future,' said Peter Gijsberts. 'With small jet engines they'll be lighter, and have a flight time of around ten minutes. But they still won't be available to the public. You won't be able to buy one in Harrods.'

'Small turbofan jet engines, like those used in cruise missiles, are available,' added Frank Dickman, 'but they are not much improved from those used in the Bell Jetbelt. They offer a considerably longer flight time, but are complex, expensive, and heavy. The Bell pilots could not carry the weight of the old jetbelt unless the engines were running.'

Rocketbelt pilot Bill Suitor was relieved that personal flying machines had not become publicly available and pleased to see rocketbelts remain as entertainment attractions. 'I hope they never become commonplace,' he said. 'Nobody would be safe. One collision over the neighbourhood, and down your chimney they would come like an unwelcome Santa Claus. Then there is the question of telephone and power wires. I know first hand how dangerous wires are. You just can't see the things. I had several close calls, and almost sliced myself up like a big ripe brick of stinky cheese.'

In the mid part of the 20th century, a generation of science fiction writers thrilled fans by creating an imaginative array of fictional technological innovations. But such is the rapid progress of technology, that many of these imaginary creations were soon bettered by real-life inventions. Mobile phones, microwave ovens, and space shuttles knocked Buck Rogers' walkie-talkies, synthetic foodstuffs, and sparkler-powered spaceships into a cocked spaceman's helmet. Science fact has out-trumped science fiction in so many ways. Mankind has managed to avoid wearing silver jumpsuits and living underground, but we have invented the internet and walked on the moon.

In the 21st century, most all of us have flown in an airplane, a helicopter, or even a hot air balloon. Thanks to centuries of obsessive invention, flying is easy. But the reality of a personal rocketbelt continues to intrigue and elude us. It is possible to buy an airline ticket and fly over mountains and oceans to the other side of the world. But it remains impossible to strap on a rocketbelt and fly over congested sidewalks and traffic jams to the local grocery store. For the time being, those who covet their own rocketbelt will have to remain with their feet firmly on the ground. Perhaps that is no bad thing.

'Can you imagine every moron who could afford one zooming all over the sky with a rocketbelt?' said Bill Suitor. 'God help us!'

BUILDING A ROCKETBELT
A cautionary essay by Frank Dickman

More than forty years ago, a daring band of engineers created one of the unforgettable icons for an era of promise. They envisioned and built the Bell SRLD – Small Rocket Lift Device, better known to the general public as the Bell Rocketbelt. If you take the time to go back to read the popular science and technology literature of the 1960s, you will quickly realise how much that decade truly was a time of creativity and promise. Plastics, computers, Teflon cookware, transistor radios, calculators, microwave ovens, hang gliders, gyrocopters, flying cars, ATVs, heart transplants, rocket programs, and men in space. By the end of the 1960s, there would be a man on the moon. It was an age of dream and promise, and nothing seemed impossible.

While the Bell Rocketbelt captured the imagination of the common man, it was never destined to be a practical device. Twenty-one seconds of fuel was its primary limitation, although there were others. The marketing department at Bell Aerosystems envisioned all sorts of potential applications. These imaginative scenarios ran the gamut from military assault to fire rescue. But only the practical dreams of the Age of Aquarius would survive to maturity.

Today the rocketbelt still inspires interest, imagination and awe. It is still the dream of many to buy, or build, or own one. On the surface, it seems to be a simple device. Wasn't it designed, after all, for the average soldier? It has always seemed somehow within the reach of the backyard mechanic. It is not. It just looks easy, the same way Olympic gymnasts and figure skaters make what they do look easy. The handful of successful builders all derived their craft from close observation of the original Bell device.

The dream of the rocketbelt, and the concept of individual flight without aircraft, is nowadays kept alive by handful of dreamers in every country in the world. Many are nostalgic, reflecting in middle age on the past promises of their youth. Some have considerable knowledge. Some do not. A very precious few have the skills and wherewithal to build their own version, and the steady nerves to try it out. They are few indeed.

The original Bell Rocketbelt was built by a large team of engineers, technicians, and craftsmen employed by a creative aerospace company. They were funded by the US government with what would be today's equivalent of $1.4 million dollars. They were also well experienced in the design of the hydrogen peroxide thrusters Bell had previously created for the top secret X-1 supersonic jet program in a successful US effort to be the first to break the sound barrier and exceed the speed of sound.

The education, skills, and experience required would have included mechanical engineering, chemical engineering, metallurgy, and jet nozzle design. Advisors from the nascent US space program and from a contingent of former Nazi rocket scientists were consulted, as well as physicians and ergonomics experts. Exotic metal welders, machinists, draftsmen, materials testing technicians, and fibreglass melding craftsmen provided specialised knowledge and skills. There were also riggers, safety spotters, cameramen, and firemen. A team of not less than 13 such experts were present for every rocketbelt test flight.

This is not a small contingent of experts, and not a simple project for a high school shop class. If it looks simple, it is not.

A good historical foundation on the subject may be found at Peter Gijsberts' rocketbelt website, www.rocketbelt.nl. Also to be found there are some links to secondary sources for copies of patents and government reports. Other useful reference sources include the *Reader's Guide to Periodical Literature* and the *Applied Science and Technology Index*, both available from some public libraries, and via the internet. Extensive web searches can also locate contemporary technical information and materials sources. A thorough technical study of the subject is essential.

The original Bell Rocketbelt is widely understood to have been constructed of components borrowed from the Mercury Space Program, or specifically developed by Air Force defence contractors. In the last 40 years much of this technology had trickled down to more common usage in laboratory and chemical plant applications, much as Velcro and Teflon are in common usage today.

Using a simple drafting program, it is possible to draw the first of several schematics called Piping and Instrumentation Diagrams (P&IDs). One purpose of such diagrams is to allow the designer to visualise a complex system, and consider the interaction of safety and control devices. What if there is a leak here? What if there is overflow there? What if there is an unexpected pressure excursion? What if a runaway reaction occurs? What might be a possible sequence of disasters and what sequence of safety devices must be triggered? As the P&ID is modified, a multitude of calculations are performed to determine the required flow rates, pressure losses, dimensional sizes, areas, volumes, and pressure ratings for each component. Materials also need to be specified for chemical compatibility, strength, weight, cost and availability, or ease of fabrication.

Engineering is the practical application of science. Often it is the art of well-considered compromise. An engineer stranded in the wilderness doesn't say, 'If only I had a million dollars and a thousand men, I could build a vehicle to fly me out of here.' Instead, an engineer looks to see what materials and resources are available, and creates a solution from the materials at hand. It's not pure science. It's knowing enough applied science to create something that works.

Once an overall plan is conceived, based on available materials, it is generally subjected to the withering crossfire of a negative peer review. In this phase, other engineers are invited to criticise any weaknesses in the design. Following peer review, specific products are selected based on function, material specification, availability and price. Usually, three products or competitive prices are compared for each item required. The lowest priced source is not always the best selection. Reputation, reliability, performance, and ease of doing business are also key factors. Compromises and redesign are common when unexpected developments occur – as when a vendor admits that while the product is pictured and described on their website and in their catalogue, they don't actually have any in stock and haven't made one for years.

After the initial P&ID is developed, the internet search can begin for miniature, lightweight, high-pressure components. Supplier sources, catalogues, and pricing need to be determined.

After specifications are reviewed, preferred sources, prices, specifications, and weights are typically detailed on an electronic spreadsheet as the first stage of a feasibility study.

The fuel is also less readily available than it was decades ago. It is now very expensive. It is primarily sold to prequalified clients, and delivered to approved facilities in tanker truckloads. Delivery charges are as high as $25 per mile, plus travel and living expenses for two factory-assigned technicians.

If purchased commercially, the fuel will cost about $1,500 for each 21-second flight of a pilot weighing 150 pounds. Bell pilots required a minimum of 20 practice flights, while tethered to safety lines. For the potential rocketbelt builder, it will therefore be necessary to figure out how to fractionally distil and purify a small quantity of H_2O_2 from scratch just for compatibility test purposes, as well as small quantity fuel requirements later.

But the information is readily available on the internet and all materials are available for purchase there. The system can be professionally built on a limited budget without the resources of an aerospace company. The most difficult obstacle is in finding suppliers willing to provide extremely small quantities of exotic materials or services at discount prices. Some of the specialty alloys required will be tough to find in small quantities.

Most rocketbelt information on the internet is in the form of stories from the popular press, where precise information is sketchy and inaccurate. A good technical starting point is the US Patent and Trademark Office (www.uspto.gov). Unfortunately, patents issued prior to 1975 cannot be searched online at the present time. They are largely expired patents and represent ancient history. But by persistence on the paper trail, it is possible to obtain copies of the original Bell patent as well as earlier and later related patents, about fifty in all. Anyone familiar with this process is probably also aware that patents are written in a confusing, anachronistic style, with illustrations that can be misleading or inaccurate, and with claims that may never be incorporated in the device as built. Many patented inventions are never built at all. Many are found to be unworkable or unmarketable.

In the early days of the space race, a number of excellent scientific reports were produced and distributed to engineering personnel in order to provide a common set of design documents to assure that everyone was working from the same page. The internet provides access to a variety of sources including the NASA Technical Report Server, and National Technical Information Service (NTIS) which maintains copies of reports on government-funded projects. Reviewing other websites, it is possible to find excellent new references on the propulsion design, and downloadable software that calculates key propulsion characteristics.

Be aware that some of the information published on this topic is erroneous, both on the internet and in print. You may come across references to two or three 'technical' building manuals on the subject published in the early 1990s, such as Derwin Beushausen's *Airwalker* study manual. They are all currently out-of-print, but copies are available if you search hard enough. Be warned, however, that these books are self-published, homemade manuals. Although the dedication and good intentions of the author are clear, they can turn out to be a quasi-informative collection of partial information by an enthusiastic but untrained author who has not built or demonstrated any of his ideas. In spite of all this, there is sometimes a lot of good general information, assembled through the contribution of others.

Some of the dimensional information in these 'technical' books is right on the mark, because it was measured directly, or as is implied in some of the text, there was help from someone who knew the subject and how to calculate. Other dimensional information is incorrect, and the authors do not agree on dimensions. At least one of the published books, while very enthusiastic and interesting, contains serious errors and omissions. These errors are both misleading and potentially dangerous. The challenging difficulty is in separating the good information from the hazardous information. Expert help is strongly advised. A rocketbelt cannot be built from a cookbook.

In general, these documents do contain helpful information. The challenge for the new reader is in separating the wheat from the chaff. It isn't easy. Doubt everything. Check everything. Measure everything twice. Newspaper reporters have a credibility motto that goes: 'If your mother says she loves you – check it out!' It is good advice.

Reading and researching the topic, like any hobby, requires time and diligence. If the hobbyist reader wishes to go further, and contemplates a construction project, there are a few things to consider. The project will need: one or more engineers (chemical, mechanical, or aerospace) with practical experience; the advice of a chemist; a first-rate machinist; an instrument technician; a certified welding technician; a pipe/tube bending technician; a valve designer; and a budget.

A $10,000 budget will barely cover material and labour on getting started, even if you possess all the skills and do most of the work yourself. It will not cover fuel, testing, and trials. As previously mentioned, Bell had a team of thirteen technicians on hand for every flight. Double or quadruple any estimated budget as a minimum and estimate three or more years of effort. Now decide how to afford it. It won't be cheap. Depending how proficient and cost conscious you might be, an estimate in the range of $50,000 to $200,000 is probably about right. For that price you could buy a Lamborghini.

For comparison, several technical universities sponsored lightweight solar powered vehicle tests in the US in 2003. Each was built by student volunteers and each cost over $100,000. The vehicles are mostly constructed of bicycle parts, a carbon fibre shell, and solar panels. Pretty expensive, even for high-performance bicycle parts.

Similarly, the reproduction engine for the airplane replica built in the US for the 100th anniversary of the Wright brother's flight cost over $100,000 for the engine alone. It is an inefficient twelve horsepower engine that can only run for one to five minutes before seizing for lack of oil (which was not circulated in the original design). This gives some idea of cost, if you must rely on work by others.

The rocketbelt fuel, a concentrated grade of peroxide bleach, has been receiving renewed attention because of its relatively benign Environmental Protection Agency (EPA) classification. The products of H_2O_2 combustion are water (as steam) and excess oxygen. Of all the rocket fuels and fuel exhausts produced in the world, this is the most benign and environmentally friendly chemical reaction. In accordance with EPA regulations, accidental fuel spills can be rinsed away with a garden hose. Spills of traditional rocket fuels, like Monomethyl Hydrazine and Nitrogen Tetroxide require area quarantine and the response of a fully equipped Hazardous Materials (HazMat) team.

Unfortunately, H_2O_2 reacts with virtually everything except a small number of materials. When 90-percent rocket grade H_2O_2 reacts, it expands to roughly 5,000 times its original volume, at standard temperature and pressure, in about one twentieth of a second. In reaction with combustible organic materials, it generates heat, and excess oxygen that can maintain a flame.

Material compatibility is an important consideration, because many common materials can react violently with Hydrogen Peroxide. Only a handful of materials are relatively inert, and even these must be professionally treated to surface passivation with hot concentrated Nitric Acid (HNO_3) to remove microscopic residues of tool steel, cutting fluid, and other impurities, before being re-tested for material compatibility. Acid passivation needs to be performed for all metal wetted surfaces in contact with H_2O_2. Surfaces have to be very clean, similar to those required for Oxygen service.

It is necessary to thoroughly understand chemical resistance and compatibility tables and charts. Many chemical equipment supply houses provide this information. It is necessary to establish a list of materials with which H_2O_2 does not react, and use those exclusively. The list must include metals, polymers, seals, and lubricants.

Safety concerns deserve consideration. Proper safety equipment for working with H_2O_2 is required. At a minimum, these include appropriate polymer coated splash-proof chemical outerwear with hood, gloves, and booties, goggles, and full faceplate, plus emergency shower and eyewash, fresh water hose with quick operating nozzle, and a backup technician for safety. Knowledge of chemical safety practices and procedures and manufacturer Material Safety Data Sheets documentation is required.

Note that the fuel control valve is a critical design component, and is not commercially available. While it functioned adequately for the application, the original design is considered poor. It was produced by a defence contractor that specialised in hydraulic aileron lifters. They were not a valve designer nor manufacturer, so their design follows the methods of which their engineers were most familiar. Their original design was unnecessarily complicated, heavy, bulky, and trouble-prone. It leaked in the closed position and exhibited too much binding friction. The fact that no commercial valves are manufactured this way today is further indication that it was a design of questionable utility.

The custom rocketbelt fuel valve cost $3,500 in 1970 (about $12,000 in today's dollars). On the bright side, numerous manufacturers currently produce small 3,000 pounds per square inch gauge (psig) mass-production valves that do not require impossible levels of machining accuracy or unusual coatings and yet provide reliability, low friction and zero leakage, internal or external. They typically cost under $300. They do require custom modification to provide the actuation and required flow characteristics for this application, but this can be done quite economically by anyone who knows how.

In addition to the myriad of safety and control devices and instrumentation previously mentioned, the rocketbelt also requires controls to maintain balance, mobility, pitch, roll and yaw. All of this has to be installed within an extremely small footprint, balanced on either side of a centreline, with no room for error.

In addition to material compatibility and passivation, knowledge of strength of materials, flow rates, and pressure tests is required.

Frequent strength of materials calculations are required throughout the design process. American Society of Mechanical Engineers pressure vessel calculations are also considered critical. A safety factor of 1.5 to 3.0 is typically recommended. Pressure tests are essential. Load tests are also useful.

Flow rates, pressures, and pressure drops (losses) need to be calculated for each system component. Every mistake will cost at least $1,000 to fix, or can harbour the potential to create a disaster.

Anyone can build a rocketbelt. All they need is an aerospace engineering degree, a dozen years of experience in chemical plant design, a few industry connections, a can-do attitude, some spare liquid assets, and a lot of free time. It's easy, only if you're already a rocket scientist.

Frank Dickman is a graduate of the Illinois Institute of Technology with a Bachelor of Science Degree in Mechanical and Aerospace Engineering. His consulting engineering experience, US and international, spans approximately 17 years. His technical marketing and product management experience spans about 15 years. A bibliography of his numerous science articles and white papers may be found on the internet. The EagleOne website can be found at: www.geocities.com/EagleAeroFlight

Note: The author of this technical commentary accepts no responsibility for the accuracy or use to which information contained within is interpreted or applied, either by other professionals or the laity. Do not attempt this at home.

ACKNOWLEDGEMENTS

Murder, kidnapping, assault, gunplay, theft, drugs, a $10 million lawsuit... It's not surprising some of those involved in this bizarre tale were keen to keep quiet about it. But many among the rocketbelt community did offer their help during the writing of this book, and their assistance proved invaluable.

Peter Gijsberts, a rocketbelt expert from the Netherlands, is the owner of exhaustive website www.rocketbelt.nl. Peter's open-handed assistance was greatly appreciated, and he also supplied a selection of photographs for these pages.

Frank Dickman, an extremely knowledgeable and experienced engineer and rocketbelt builder, provided much useful information and kindly wrote the excellent *Building a Rocketbelt* essay exclusively for this book. I am very grateful to him.

Kathleen Lennon Clough organised the International Rocketbelt Convention, and she graciously provided information and photographs relating to the event. Rocketbelt expert Ron Carey was also generous in supplying useful information, including archive documents and video footage.

When it came to uncovering the truth behind the Rocketbelt 2000 caper, tracking down those most closely involved was often particularly difficult.

When I first caught up with Brad Barker he was enjoying a chicken fried steak in a Houston restaurant. He was just as friendly and charming as those involved with him had reported, and generously spoke about the rocketbelt caper. 'It's been quite a deal, I'll tell you that much,' he said. Barker's attorney, Jeff Haynes, was also very helpful.

Larry Stanley was easier to locate, but less willing to talk about the Rocketbelt 2000 caper. I wrote to him in prison but, after exchanging several letters, he decided against being interviewed. As a result, his side of the story has been ascertained from his public affidavits and testimonies provided during the various rocketbelt court cases. In addition, Tommy Mackey, Stanley's brother-in-law, and Michael Von Blon, Stanley's attorney, were kind enough to keep me updated on developments.

Of course Joe Wright is no longer around to tell his side of the story. Nancy Wright, Joe's younger sister, was able to offer a unique insight into his life via her contact with his friends and the rest of his family. 'His family were very proud of him,' she said. Maurice Heimlich, a close friend of Joe Wright, was also kind enough to supply his memories.

Bill Suitor, is regarded as the ultimate rocketbelt pilot, although the man himself insisted, 'I don't know if I'd call myself the ultimate anything.' Now retired, he was travelling around the US West Coast sightseeing and selling carvings of wildlife when I caught up with him. His detailed memories of flying the Bell and Tyler Rocketbelts, and the Rocketbelt 2000, proved invaluable. Bill emailed me regularly and also provided me with photographs from his personal collection for publication in these pages.

I am also grateful to engineer Doug Malewicki for his assistance, and to Danny Carr from Nobles Gate film production company for swapping information and providing his photograph of Brad Barker that appears in these pages. I'd also like to thank Powerhouse Director of Operations Dan Schlund, rocketbelt builders Stuart Ross and Juan Manuel Lozano, and to the many rocketbelt enthusiasts with whom I corresponded over the course of writing.

Final thanks go to all who read, reviewed, supported and assisted the book during its creation. Keep reaching for the skies.

Paul Brown first wrote about the rocketbelt caper in the July 2003 issue of JACK magazine. This is his fourth non-fiction book. He lives near Newcastle upon Tyne in the North East of England.